MATEWAN
BEFORE THE MASSACRE

WEST VIRGINIA AND APPALACHIA

A SERIES EDITED BY RONALD L. LEWIS

VOLUME 8

OTHER BOOKS IN THE SERIES

Transnational West Virginia
Edited by Ken Fones-Wolf
ISBN 0-937058-67-x [HC] • ISBN 0-937058-76-9 [PB]

The Blackwater Chronicle
Edited by Timothy Sweet
ISBN 0-937058-65-3 [HC] • ISBN 0-937058-66-1 [PB]

Clash of Loyalties
By John Shaffer
ISBN 0-937058-73-4 [HC ONLY]

Afflicting the Comfortable
By Thomas F. Stafford
ISBN 1-933202-04-1 [HC ONLY]

Bringing Down the Mountains
By Shirley Stewart Burns
ISBN 978-1-933202-17-4 [PB ONLY]

*Monongah: The Tragic Story
of the 1907 Monongah Mine Disaster*
By Davitt McAteer
ISBN 978-1-933202-23-7 [HC ONLY]

MATEWAN
BEFORE THE MASSACRE

POLITICS, COAL, AND THE ROOTS OF CONFLICT
IN A WEST VIRGINIA MINING COMMUNITY

MORGANTOWN 2008

West Virginia University Press, Morgantown 26506
© 2008 by West Virginia University Press

First edition published 2008 by West Virginia University Press
Printed in the United States of America

15 14 13 12 11 10 09 08   9 8 7 6 5 4 3 2 1

ISBN-10 1-933202-28-9
ISBN-13 978-1-933202-28-0
(alk. paper)

Library of Congress Cataloguing-in-Publication Data

Matewan before the Massacre. Politics, Coal, and the Roots of Conflict in a West
Virginia Mining Community / by Rebecca J. Bailey.
p.   cm.
(West Virginia and Appalachia; 8)
1.Coal mines and mining. 2 Coal mines and mining – West Virginia. 3. West
Virginia Mine Wars, W. Va., 1897–1921. Strikes and lockouts – Coal mining
– West Virginia – Kanawha County – History. 4. West Virginia – Matewan –
History. I. Title. II. Bailey, Rebecca J. III. Series.
IN PROCESS

Library of Congress Control Number: 2008936435

Cover Design by Than Saffel
Cover Photographs: Matewan Station, Matewan, West Virginia ca. 1920. Photo-
grapher unknown. Norfolk & Western Historical Photograph Collection. Used by
permission of Norfolk Southern Corporation.
Printed in USA by United Book Press

For
William Lee "Bill" Thompson
(1969–2001)

"I still miss someone. . . ."
—Johnny Cash

CONTENTS

ACKNOWLEDGMENTS

ONE VERY BAD DAY while in graduate school, I sat on the side steps of Woodburn Hall in Morgantown West Virginia, feeling so low that when a student I called "Big John" walked up and asked, "Ms. Bailey, you look like somebody's died, what's wrong?" I looked up at him and said "John, have you ever felt so bad about yourself, you wonder why you're using up somebody else's air?" John just looked at me and said, "Well, I could sit here and tell you how great I and the other students think you are, but I'm going to tell you something somebody told me once that helped me a lot." That really piqued my curiosity because John was a Vietnam Vet, a recovering alcoholic and drug addict, and every second of his hard life seemed etched in the lines of his weather-beaten face. What John proceeded to tell me was that when I was down on myself all I had to do was imagine my life as a kingdom with inhabitants I had brought in—and then ask myself, "who lives in my kingdom? Have I peopled my life with scoundrels, thieves, and knaves, or queens and princes?" Then John looked down at me and said "Ms. Bailey, are you going to tell me that you've surrounded yourself with scoundrels, thieves, and knaves?" And I said "No, I've been blessed all my life to be surrounded by people who've wanted and done only the best for me." To which John replied simply, "To know your own measure, all you have to do is accept that the inhabitants of your kingdom are a reflection of you." Self-pity instantly turned to relief and gratitude—how could I be worthless if so many wonderful, admirable people cared for me?

This story was the best way I could think of to lead into this acknowledgment of all who have brought me to this day. I have been buoyed up time and again by the love, faith, and strength of so many. Throughout my life I have been loved by friends without parallel: Lia, Toni, Kim, Alec, Leslie,

Eddie, Tammy, Sara, Stephanie, Denny, Chris, Connie, Carletta, Shirley, Rhonda, Delilah, and Diane. When I became a teacher, I finally understood what a challenge (and, I hope, occasionally a source of satisfaction) I must have been to Mr. Strathairn, Mrs. Mitchell, Mr. Keys, Mr. Lloyd, Mr. Wilson, Mrs. Pendergrass, Ms. Walker and Drs. Chappell, Canning, and Crapol. Once I started graduate school and gave my life over to the study of Matewan, the following people provided invaluable inspiration and support: Mary Lou Lustig, Jack Hammersmith, Jack McKivigan, Liz and Ken Fones-Wolf, Teresa Statler-Keener, Mike Ruddon, Jeff Cook, John Hennen, Paul Rakes, Paul Yandle, Lou Martin, Ken Sullivan, Stuart McGehee, Phil Obermiller, Shannon Wilson, Harold Forbes, Christy Venham and the staffs of the Eastern Regional Coal Archives, the West Virginia Division of Culture and History, and, most especially, the past and current staffers of the West Virginia and Regional History Collection. I can never say enough about all of the folks at the WVU Press who went above and beyond too; eternal gratitude to Pat, Than, Connie, and Rachel. Since completing my dissertation, supportive colleagues and friends have seen me through this last leg: Ann McCleary, Steve Goodson, John Ferling, Elmira Eidson, Sandra Stone, Pete Crow, Dan Williams, Susan Mead, Cece Conway, Lauri Anderson, and Karl Precoda. Special thanks are owed to my colleagues at Northern Kentucky University, most especially: Eric Jackson, Carol Medlicott, Debra Meyers, and Jeffrey Williams. To her probably eternal chagrin, I credit my career as a public historian and public history educator to Dr. Barbara J. Howe. What research skills I possess, I owe to her. Moreover, the day I entered Barb's Introduction to Public History class, I found the path to a professional life full of wonder, where I could lend my intellect not only to the celebration of history, but also to the creation and dissemination of history that includes and empowers.

At the feet of my last teacher and mentor, Dr. Ronald L. Lewis, lies the responsibility for the work within. Until he asked if I had considered staying at West Virginia University to work on a Ph.D., which would allow me to be the graduate student to write about the oral histories John Hennen and I collected in Matewan, I had fully intended to be a footloose public historian. Not long after that, when Ron said that if I could explain why the Massacre happened in Matewan he could guarantee me a book contract the next day, I was hooked. It only took me eleven years to finish the Ph.D., and seven more to complete the book. Throughout it all, Ron coached, guided, goaded, and waited, good-naturedly accepting my threats to dig him up and forge his signature if he died before I finished. But through it all what mattered most was when he asked how I knew I was right about something and I'd answer that it was what my coal miner grandfather would have concluded, he would just nod—he was a son of coal himself and he knew what kept me going was deeper than academic or professional ambition or curiosity. I was compelled by my blood, my identity, and my people.

Thus, I turn last to the people I owe the most and whose opinions matter most. To the people of Matewan, Mingo County, and southern West Virginia, especially Margaret Casey, Margaret Hatfield, Robert Ward McCoy, Paul McAlister, and Yvonne DeHart, I did my best and I hope you all are not disappointed in me. To my siblings, nieces, and nephews, this is what I was doing when I wasn't being a better sister or auntie. Please know how much the fact that you loved me anyway will live forever in my heart. To my Thompson family, you loved me as one of your own and dreamed for me too—my love always to you. And last and most, my alpha and omega, Carl and Phyllis Bailey—the same courage, hope, strength, and will that propelled you far from home led me back and saw me through to this day. You raised me right; I hope it shows.

MAY 19, 1920

"Not one of the writers who went into Mingo . . . ever wrote the true story."
– *Mary Harris "Mother" Jones*

WEDNESDAY, MAY 19, 1920, dawned dreary and overcast. Though rain drizzled from the clouds intermittently throughout the day, the small town of Matewan in Mingo County, West Virginia, teemed with miners, as union relief funds were being distributed. In the midst of the activity, at 11:47 a.m., a party of Baldwin-Felts agents disembarked from train #29, having come to Matewan to enforce eviction notices for the Stone Mountain Coal Corporation. According to Walter Anderson, one of the surviving agents, Albert Felts, contacted Mingo County Sheriff G. T. Blankenship seeking his help in processing the evictions. He was denied, but managed to secure authorization from a local justice of the peace. Chief of Police Sid Hatfield and Mayor Cabell Testerman confronted the Baldwin-Felts agents as they made their way through town. Both Hatfield and Testerman contested the agents' authority to process the Stone Mountain evictions. Their primary arguing point, Hatfield later asserted, was that because the houses in question lay within Matewan's municipal limits, he and Testerman possessed the jurisdictional sovereignty to halt the evictions.[1]

At this point, the confrontation ended and the Baldwin-Felts agents crossed the railroad tracks and proceeded up Warm Hollow to process the evictions. Hatfield and Testerman left to telephone county officials. Testerman allegedly called Mingo's prosecuting attorney Wade Bronson to enquire about the legality of the evictions. After Bronson read him "the Red Man's Act—the riot act," Testerman authorized Hugh Combs, a Methodist "exhorter" and local miner, "to obtain reliable men to protect the town." One of Matewan's two telephone operators later testified that Hatfield told either Blankenship or Deputy Sheriff "Toney" Webb that "those sonsabitches will never leave here alive." Throughout the afternoon, armed men arrived in Matewan, and the town became "a powder keg." The situation grew so tense that Matewan Grade School let out early, and the children were sent home to get them off the streets.[2]

Hatfield, Testerman, and a crowd of miners, men, women, and children spent the afternoon watching the Baldwin-Felts agents carry out the Stone Mountain evictions. At one point, Hatfield allegedly approached Albert Felts, who raised his gun and told Hatfield he was trespassing on private property. Smiling, Hatfield replied, "That's alright, I'm a private man," and continued his advance on Felts. When Hatfield drew near, Felts observed that at Paint Creek he had been shot at from ambush but had refused to be "bluffed out." Hatfield assured Felts that if there were any trouble, "here, no one will go to the hills on you . . . . They will come face to face." Testerman again asked Felts to desist. Felts refused but offered to stop and return to town if Testerman could prove that he was acting illegally.[3]

At approximately 3:30 p.m., the Baldwin-Felts agents completed their work, came back across the tracks into Matewan, and checked into the Urias Hotel. Although miners and Police Chief Hatfield had observed the six evictions, there had been no more confrontations. A surviving Baldwin-

Union relief day in Matewan. *Miners from tent colonies line up in Matewan to receive union relief funds and food supplies. On the morning of May 19th, 1920, union relief was distributed, swelling the number of people in the town when the Baldwin-Felts agents arrived. [West Virginia Regional History Collection/WVU Libraries]*

Matewan Street Scene, 1920. *Because of the town's position between the river and the railroad, the buildings to the right here had "double fronts." One front entrance faced the railroad, where passengers disembarked, and the other opened onto Main, or Mate Street, shown here.* [WVRHC]

Felts agent later recalled that the proceedings had gone smoothly, citing as proof the agents' transfer of one family's belongings to another location at the evictee's request. In contrast, local recollections of the evictions present the agents' actions as the primary stimulus for the gun battle. The agents allegedly arrived heavily armed and proceeded to bully everyone they encountered, callously and haphazardly piling the belongings of a miner's sick wife in the rain, for example. However the agents comported themselves, they made their way to the Urias unmolested.[4]

Once at the hotel, the agents disassembled and repacked their large firearms, a legal necessity since only three of them (the Felts brothers and C. B. Cunningham) possessed the required license to carry pistols in Matewan. Several individuals reported to Albert Felts that trouble was brewing and that armed miners were milling about the town. Felts gathered his men and told them that if a conflict erupted, they were not to fight or resist arrest but to go quietly because bail would be posted and the situation resolved peacefully. After repacking their weapons, all but one of the agents sat down to a meal. As time approached for the 5:15 p.m. train to Welch, the agents thanked the hotel manager, Anderson "Ance" Hatfield, for his hospitality, again brushed aside concern for their safety, and made their way across Mate Street to the railroad depot.[5]

3

Matewan Chief of Police Sid Hatfield. *"Miners' Hero" Sid Hatfield was not considered a legitimate Hatfield by many Hatfields. Opinions of his role in the Massacre served as a barometer of one's sympathies for decades. [WVRHC]*

While the agents waited for the train, Sid Hatfield approached Albert Felts and requested that he accompany him to a meeting with Mayor Testerman. Standing just inside the doorway of the Chambers's Hardware store, Hatfield, Testerman, and Felts again began to argue. Hatfield threatened Felts with arrest, to which Felts responded that he too possessed a warrant, for Hatfield. Mayor Testerman offered to post bond, saying that, as police chief, Hatfield was needed in Matewan. Felts demurred, and Testerman asked to see the warrant papers. After reading them, he declared the papers "bogus," whereupon (depending on the witness) either Hatfield or Felts pulled a gun and fired.[6]

Within seconds, a blaze of gunfire erupted. Mayor Testerman staggered away clutching his stomach while Albert Felts fell where he stood, mortally wounded. Pandemonium ensued; African-American laundry owner John Brown, who was standing at the depot waiting on a shipment, led state senator M. Z. White and his wife, who were awaiting the train, to safety in a nearby basement. According to one child witness, a thousand shots rang out in less than ten minutes. The only other armed Baldwin-Felts agents, C. B. Cunningham and Lee Felts, who were standing nearby, drew their guns, but neither made it to Albert Felts's side. The other agents scattered, seeking cover. One agent, as he ran past Mary Brown, who was looking for her husband John, asked, "What's the best way to get out of this town?" Pointing to the river, she shouted, "Split the creek!" Two agents (the wounded Anderson brothers) managed to climb aboard the waiting train before it quickly pulled out of the station. Another agent, who later claimed to have hidden in a coal shed, slipped out of town undiscovered. The other five agents received no mercy. J. W. Ferguson fled wounded to the back porch of Mary Duty's home and begged her to hide him. On the verge of hysteria herself, Mrs. Duty fled back into her house as armed men approached. She later claimed to have heard Ferguson say, "Gentlemen, I have not fired a shot in your town." Despite his pleas, Ferguson was shot again, allegedly by Fred Burgraff, who told him, "You S.O.B., you've gone too far. You're going to die."[7]

As quickly as it had started, the Matewan "Massacre" ended. Several men loaded Mayor Testerman on a train bound for Welch, where he died later that night, his only words being "Why did they shoot me? I can't see why they shot me." The bodies of the dead agents lay where they had

5

The dead Baldwin-Felts agents. *Shown here are the seven Baldwin-Felts agents killed in Matewan on May 19th, 1920, including Albert and Lee Felts, younger brothers of Agency Chief Thomas Felts. [WVRHC]*

fallen until Sheriff Blankenship, accompanied by Williamson Mayor W. O. Porter, arrived from Williamson at 7:15 p.m. Mayor Porter supervised placing the corpses onto a Williamson-bound train, which left the men who undertook the grisly task "literally covered with blood." For the rest of the night, the armed men of Matewan patrolled the town, tensely watching the trains as they sped by on the way to Williamson. Unbeknownst to them, one train carried a contingent of West Virginia state police, who thought it best to arrive in Matewan the following morning.[8]

The basic facts of the Matewan Massacre and the events it set in motion are indisputable. It occurred in the spring that followed the abortive 1919 March on Logan and just four months after John L. Lewis, the new head of the United Mine Workers of America, had stood in Bluefield, West Virginia, declaring the launch of the union's latest attempt to organize southern West Virginia. After the May 19 gun battle, which left ten men dead and unknown others wounded, the Operators' Association of the Williamson-Thacker coalfield initiated a lockout. The Matewan Massacre would become a pivotal episode in West Virginia's second "mine war." Although the Massacre was a brief but bloody exchange, the events that followed in its wake included a 28-month strike that led to two dozen deaths, West Virginia's longest and most controversial murder trial to date, a United States Senate investigation, the retaliatory assassination of Sid Hatfield, and the largest armed civilian insurrection since the Civil War. Because the Massacre also occurred in the heart of the "stompin' grounds" of the Hatfields and McCoys, it entered the annals of American history as one of the best-known incidents of West Virginia and Appalachian history. What would be lost was why both the Feud and the Massacre happened in the same community separated by only forty years—and, more pointedly, why the Massacre had happened in Matewan.[9]

Within 24 hours of the events of May 19, 1920, news of the Matewan gun battle spread across the nation. Journalistic reports appeared not only in local and regional newspapers, but even in the *New York Times*. In less than two weeks, an account of the incident and its underlying causes appeared in the *Nation*, one of America's leading current affairs magazines. That first story, entitled "Private Ownership of Public Officials," set the tone and defined the parameters of the contemporary record of the story.[10]

According to the United Mine Workers of America, reform-minded journalists, and liberal politicians of the day, the violence of May 19, 1920, resulted from the coal operators' attempt to defeat the UMWA's effort to organize the miners of Mingo County. The principal agents of the story thus became the labor union and its corporate foes, the anti-union coal operators of southern West Virginia. The critical issue was seen to be the right of coal miners to organize and defend their rights against the overwhelming force of "industrialism gone mad." The coal industry had enslaved southern West Virginia's miners in the "worst economic serfdom in America." Moreover, the coal operators were said to "own" the political and legal systems at the local and state levels, further depriving their employees of nonviolent redress for their grievances. Worst of all, it was pointed out, were the living conditions imposed on the miners by the coal operators. The miners lived in isolated, company-owned communities, policed and brutalized by armed guards, "private gunmen" who answered only to the operators. With such coercive power arrayed against them, the narrative concluded, it was little wonder that the "ignorant, primitive mountaineers" reverted "to their ancient way of settling trouble" by taking down their guns and killing their oppressors.[11]

Most of the journalists who covered the Massacre and subsequent events structured their accounts to titillate their readers and to inspire support for industrial reform. Details of the story were simplified, the causes of the conflict were generalized and broadly drawn, and individual players in the drama were reduced to archetypal caricatures. At the same time, issues that would complicate or muddy the picture were ignored. Matewan and Mingo County were indistinguishable spots in the barren landscape of an isolated and dysfunctional industrial empire. Sid Hatfield became a mountain Gabriel avenging the demoralized miners' brutalization. The detectives and the coal operators became the heartless tools of a remote and uncaring corporate Leviathan. As one journalist later observed in his memoir, "In the public mind, fed by newspaper emphasis on the sensational and with so little reporting of the background and personalities of [the] strike, the Matewan massacre . . . meant merely one more example of union violence. It was not as simple as that."[12]

If a factual accounting of the Massacre suffered because the contemporary journalists constructed a narrative to suit a political and economic reform agenda, the subsequent versions of the story written by academics were weakened by disciplinary bias. The first study, A. F. Hinrichs's 1923 dissertation, *The United Mine Workers of America and the Non-Union Coal fields*, reached the same conclusion as the journalists—that in the struggle between the United Mine Workers of America and industrial capitalism, what ultimately mattered was that the systemic causes of inefficiency in the coal industry threatened America's economic security and, by extension, the American public. Because Hinrichs was a typical early-twentieth-century social scientist, what did not appear in his analysis, despite his own firsthand observations in southern West Virginia, were accounts of local conditions. The people and communities actually caught up in the monumental struggle between labor and capital were little more than aggregate statistics in his account.[13]

In the decades that followed, the story of Matewan and the Massacre became the domain of labor historians. As a result, the narrative was subsumed into a conceptual model that reduced the southern West Virginia coalfields into one system in which "the miners" struggled for freedom in a region so dominated by the coal companies that their liberation required the intervention of the United Mine Workers. The titles of the earliest labor studies, which were mid-twentieth-century masters' theses, reflect their "institutional" focus on assessing the success or failure of the UMWA's effort: "Two Periods of Crisis in Labor Management Relations in the West Virginia Coal Fields, 1912–1913 and 1919–1922" (1946); "Strikes in the Southern West Virginia Coal Fields, 1912–1922" (1949); and "The History of the United Mine Workers in West Virginia, 1920–1945" (1950). Not unlike Hinrichs's study a generation before, the mid-twentieth-century analyses treated the people and communities of southern West Virginia as interchangeable pawns. In their paradigm, Don Chafin, the sheriff of Logan County, was *the sheriff*, and the coercive conditions that prevailed in the highly capitalized company town systems of McDowell and Logan were assumed to be ubiquitous throughout the region.

The New Left historians' emphasis on the social dimension of conflicts involving class, culture, and power ostensibly influenced late-twentieth-century labor historians to reconfigure the scholarly discussion of West Virginia's mine wars. David Corbin's 1981 *Life, Work, and Rebellion: Southern West Virginia's Miners, 1880–1922* typified the new approach. In contrast to the previous generation's scholars, Corbin addressed the influence of issues such as religion, education, and community life on the unionization struggle. *Life, Work, and Rebellion* offered the first reconstruction of the "miners' world" before and during the mine wars. Among the causes of the miners' rebellions that Corbin identified were the cumulative effect of the abrupt and dislocating influence of rapid industrialization and the oppressive atmosphere of the company system. For a quarter of a century, Corbin's definition of the southern West Virginia miners' world and the mine wars period has stood uncontested.

What remained unexamined throughout the twentieth century was that the conceptual underpinning, or the equation, that lay beneath the previous narratives of the Massacre and mine wars was the same. Visualize,

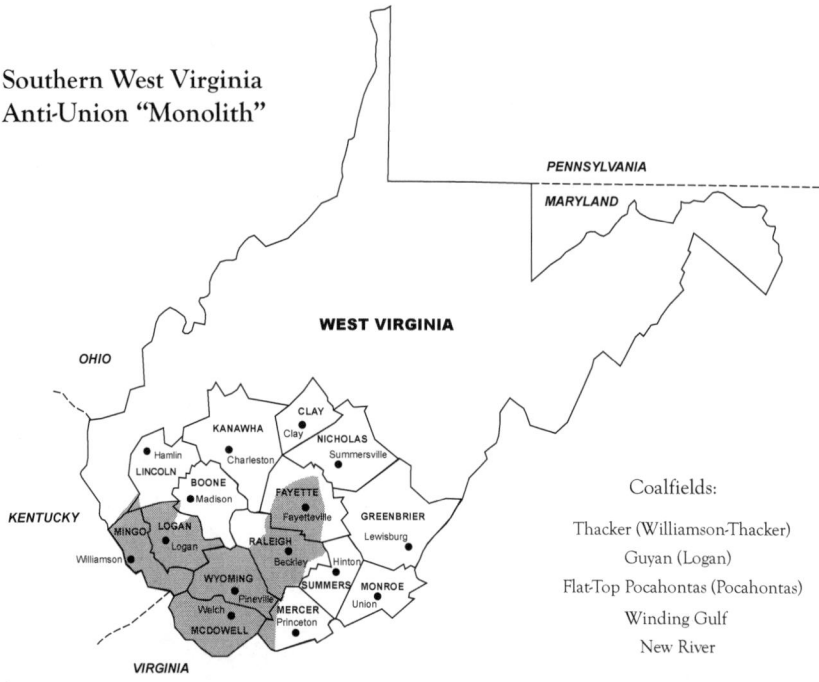

Map Intro.1 The southern West Virginia antiunion monolith. *The "monolith" roughly corresponded to Logan, Mingo, McDowell, Mercer, Fayette, Wyoming, and Raleigh counties. [Created by E-S Bonita Bonar]*

if you will, a two-dimensional line drawing representing southern West Virginia in which rest modular components representing the company/operators, the miners, and the union. The contemporary observer journalists and scholars, the mid-twentieth-century "institutional" labor historians, even the late-twentieth-century labor historians influenced by the New Left all viewed the mine wars as a chain of events that played out within this system. In their view, southern West Virginia was utterly controlled by and operated in the name of the coal companies; the miners' only avenue to liberation was rebellion, and their only ally was the union. Conclusions were drawn from a generalized body of evidence; if a fact from any one of the coal counties fit the equation, in it went. Discussion of factual complexities and historical ambiguities were surrendered to the paradigm.

Thus, *Matewan Before the Massacre* has two objectives. First, it offers the reader a "from the bottom up" examination and analysis of what caused the Matewan Massacre from the perspective of the people of Matewan and Mingo County, based on the documentary record they left behind. The following questions represent a brief selection of what this study resolves from what went unexamined, even avoided, for nearly 90 years.

First, if what had radicalized miners to the point of a violent uprising was in fact the abject demoralization of company-town life, why did the Massacre occur in Matewan, which was an independent community, rather than in a company town such as Holden in Logan County or Gary in McDowell County? Second, if the coal companies really did "own" the local political and legal systems, how did a coal miner get elected mayor of Matewan; and how and why did the pro-union Mingo County sheriff G. T. Blankenship come to power and defend the miners during his term in office? And last, why, after the failed 1919 March on Logan, did the UMWA turn its attention to Mingo County if its larger, more powerful neighbors, Logan's Guyan Field and McDowell's Pocahontas Field, were the union's true targets?

By answering these and other questions, *Matewan Before the Massacre* explains why the events of May 19, 1920, could only have happened in Matewan, in Mingo County. Its narrative reveals that the Massacre occurred in Matewan because the unique history of Matewan and Mingo County gave rise to it. Mingo County was the Achilles' heel of the corporate Leviathan that was alleged to have held absolute sway over the politics, power, and economy of southern West Virginia. The Massacre blazed out on the streets of Matewan that late spring afternoon because the miners' union seized the opportunity Mingo represented to establish a beachhead from which to organize their true targets of Logan and McDowell. By exposing the Massacre's unique origins, *Matewan Before the Massacre* explodes the grip of the anti-union monolith paradigm of southern West Virginia by demonstrating that one simply cannot use historical data from Logan or McDowell counties to explain the unfolding events in Mingo.

Matewan Before the Massacre's second objective is to recontextualize the Massacre narrative. Freed from a labor-historiography approach described by E. P. Thompson, Herbert Gutman, and Peter Friedlander as "anatomised," "essentialized," and "narrowly rational," the Matewan story will fit into a broadening historiography within Appalachian history that itself is escaping isolation and finding a place within national and international contexts. The studies that most influenced the methodological and interpretive approach taken here share an approach that deconstructs received knowledge from previous scholarship, embraces local study as a means of reexamining broader historical definition, and situates their analyses in a more nuanced holistic systems analysis. Inspired by Altina Waller's utter breakdown of the myth and factual redefinition of the participants, motives, and causes of the Hatfield-McCoy feud in *Feud: Hatfields, McCoys and Social Change in Appalachia, 1860–1900*, *Matewan Before the Massacre* challenges the reader to see southern West Virginia as a more complex political, economic, and social system than previously thought. *Matewan Before the Massacre* also borrows heavily from *The Road to Poverty: The Making of Wealth and Hardship in Appalachia* by Dwight Billings and Kathleen Blee, the case study of a single community in Kentucky that finally corrected decades' worth of damage in Appalachian scholarship caused by inadequate investigation of contextual and historical patterns. Finally, from Ronald Lewis's *Transforming the Appalachian Countryside: Railroads, Deforestation, and Social Change in West Virginia, 1880–1920* comes an understanding of Immanuel Wallerstein's world systems analysis of political economy that gives dimension and historical significance to the statement that the Massacre happened in Matewan because it was different from the other communities in southern West Virginia.[14]

A close examination of Mingo County's origins and development thus transforms the narrative of the Massacre. The abstracted paradigm of non-union southern West Virginia is replaced with a reconfiguration of Mingo as a "periphery" of the more important "semi-periphery" or even "core" areas of Logan and McDowell. Almost from its birth, Mingo was not important enough or proved too politically difficult to warrant the expense of "owning" it. Developed in the midst of the nation's second worst depression, Mingo's coalfield never lived up to the promise of its older or younger neighbors. Mingo was a place where traditional clientelism, modern welfare capitalism, and old and new social beliefs and behaviors still grappled for the upper hand. For the quarter-century before the Massacre, it was struggle, not oppression that dominated the political economy of Matewan and Mingo County. The violence that spewed forth on May 19, 1920, erupted not only from the pent-up frustrations of miners, it also vented on-going local political clashes that tapped into religious and class differences. It was a reckoning that started a war.

A special note about sources: Given the place of prominence afforded primary documents in *Matewan Before the Massacre,* the author feels that she should acknowledge what was at her disposal that had been unavailable to previous scholars. During the 1990s, partial records for the Baldwin-Felts Detective Agency and partial transcriptions relating to the Massacre and C. E. Lively trials resurfaced. Also, beginning in 1988, oral history interviews were gathered under the auspices of the Matewan Development Center, and during projects funded by the West Virginia Humanities Council in the summers of 1989 and 1990, more than 90 interviews were recorded.

NOTES

1   The time of the Baldwin-Felts agents' arrival was noted in the *Charleston Gazette,* May 21, 1920. For a sample of the variety of accounts of the Massacre, see also: *Williamson Daily News,* May 20, 1920; *West Virginia Federationist,* May 20, 1920; "Testimony of Sid Hatfield" U.S. Congress, Senate, Committee on Education of Labor, *West Virginia Coal Fields: Hearings . . . to investigate the recent acts of violence in the coal fields of West Virginia and adjacent territory and the causes which led to the conditions which now exist in said territory* (Washington: GPO, 1921), 205–221; For various eyewitness testimonies from the "Massacre trial," see *State of West Virginia v. Sid Hatfield, et al.,* H. C. Lewis Collection, Eastern Regional Coal Archives, Craft Memorial Library, Bluefield, West Virginia. (The Lewis Collection is comprised of the papers of the Baldwin-Felts Detective Agency founder Thomas L. Felts and contains the only known copies of trial records from the Matewan Massacre Trial (January–March 1921) and the trial (October 1921) of C. E. Lively, William Salter, and George Pence for the murders of Sid Hatfield and Ed Chambers.) See also "Trial Testimony of Hugh Combs," Lewis Collection, ERCA; "Testimony of Sid Hatfield," *West Virginia Coal Fields,* 219; Telegram of Walter Anderson to John J. Cornwell, May 20, 1920, John J. Cornwell Papers, West Virginia Regional History Collection, West Virginia University, Morgantown, West Virginia; The two justices of the peace for Magnolia District in 1920 were R. M. Stafford of Thacker and A. B. Hatfield of Matewan; 1920 *West*

Virginia Legislative Handbook and Manual, 763; "Testimony of Sid Hatfield," *West Virginia Coal Fields*, 206; The intersection of "civil behavior" with other stresses in tense situations is a phenomenon often discussed in studies of "domination and resistance." On this subject, see: Jane Dailey, "Deference and Violence in the Postbellum South: Manners and Massacres in Danville, Virginia," *Journal of Southern History* 63 (August 1997): 553–590, 556.

2   Sources claiming the warrants were legal: "Statement of Mr. E. C. Price," excerpt from Report of #9, May 29, 1920, Lewis Collection, ERCA. Sources claiming the warrants were not legal: "Trial Testimony of Jesse P. "Toney" Webb," Lewis Collection, ERCA; Source for deputizing of Combs and "reliable" men: "Trial Testimony of Hugh Combs," Lewis Collection, ERCA; Passed in 1882, the "Red Man's Act" criminalized conspiracies between two or more men to steal, damage, or destroy another's property; it also allowed any individual involved in the conspiracy to be tried for first degree murder if anyone died during the commission of a conspiratorial act. See Sections 10 and 13 of Chapter 35 "An Act Concerning Deadly Weapons, Etc," in *Acts of the West Virginia Legislature for the year 1882* (Wheeling, WV: W. J. Johnston, Public Printer, 1882), 421–424; *McDowell Recorder*, March 4, 1921; Query by defense attorney, "Testimony of Jesse P. "Toney" Webb," Lewis Collection, ERCA; Hawthorne Burgraff interview and Venchie Morrell interview(s), Summer 1989 and Summer 1990 Matewan Oral History Project.

3   Source for exchange between Hatfield and Felts: "Trial Testimony of Dan Chambers" and "Statement of Miss Jennie Mullens," August 21, 1920, Lewis Collection, ERCA; Rhodri Jeffreys-Jones, *Violence and Reform in American History* (New York: Franklin Watts, 1978), 91; "Trial Testimony of Mrs. Elizabeth Barrett," Lewis Collection, ERCA.

4   Source for eviction activities: Walter Anderson telegram, Cornwell Papers, WVRHC; Source for how townspeople perceived evictions: Hawthorne Burgraff interview and Venchie Morrell interview (1990); Source for Kelly's wife was the sick woman: "Trial Testimony of Charlie Kelly," Lewis Collection, ERCA; Source for agents unmolested: Dudley Williams statement, Lewis Collection, WVRHC.

5   Source for firearms break down: Walter Anderson telegram, Cornwell Papers, WVRHC; Source for: agents warned, if trouble, brushed aside: Statement of Miss Jennie Mullens, August 21, 1920, Lewis Collection, ERCA; Source for agents

warned: "Trial Testimony of Joe C. Jack," February 25, 1921, unknown newspaper, Matewan Omnibus Collection, ERCA; Source for meal: *Charleston Gazette*, May 21, 1920. The Anderson Hatfield who ran the Urias Hotel was called "Ance" to differentiate him from his more famous kinsman Anderson "Devil Anse" Hatfield.

6   Ernest Hatfield interview with John Hennen, Summer 1989 Matewan Oral History Project; "Trial Testimony of C.E. Lively," February 25, 1921, unknown newspaper, Matewan Omnibus Collection, ERCA; Sources for location of argument: "Trial Testimony of Joe C. Jack," February 25, 1921, unknown newspaper, Matewan Omnibus Collection, ERCA; "Testimony of Sid Hatfield," *West Virginia Coal Fields*, 206; Ernest Hatfield interview; Source for who said Felts fired first: G. T. Blankenship, *Charleston Gazette*, May 21, 1920; Hawthorne Burgraff interview; Source for who said Sid fired first: twelve Matewan Massacre trial witnesses stated that Sid Hatfield shot first, "Closing Argument of Wade Bronson," March 19, 1921, unknown newspaper, Matewan Omnibus Collection, ERCA; John McCoy interview with John Hennen, Summer 1989 Matewan Oral History Project; Source for who said Isaac Brewer (an uncle by marriage to Ed Chambers) fired first: "Smokey" Mose Adkins interview with John Hennen, Summer 1989 Matewan Oral History Project.

7   "Trial Testimony of Joe C. Jack," February 25, 1921, unknown newspaper, Matewan Omnibus Collection, ERCA; Source for "this town" and "split": Johnny Fullen interview with Rebecca J. Bailey, Summer 1990 Matewan Oral History Project. Mr. Fullen did not specify whether "Senator White" was Hiram S. White or M. Z. White, both of whom had served as state senators. But given that M. Z. White had been in Charleston just the day before, it was likely that he, and not H. S. White, had nearly been caught in the crossfire; Dixie Accord interview with John Hennen, Summer 1989 Matewan Oral History Project; Howard Lee, *Bloodletting in Appalachia: The Story of West Virginia's Four Major Mine Wars and Other Thrilling Incidents of Its Coal Fields* (Morgantown, WV: West Virginia University Library, 1969), 54; *Charleston Gazette*, May 21, 1920; Ruby Aliff interview with Rebecca J. Bailey, Summer 1989 Matewan Oral History Project; "Statement of Mrs. Billy (Mary) Duty," in September 9, 1920 Report of #19, Lewis Collection, ERCA; Alleged statement by Fred Burgraff, "Trial Testimony of C. E. Lively," February 25, 1921, unknown newspaper, Matewan Omnibus Collection, ERCA.

8   Source for Testerman's last words: Notes taken (probably by T. L. Felts)

following meeting with owner of the *Mingo Republican*, Dr. R. M. Musick, May 22, 1920, Lewis Collection, ERCA; "Slight Flareup," *Charleston Gazette*, May 21, 1920; "Trial Testimony of Jack Gallion," Lewis Collection, ERCA; J. W. Weir to John J. Cornwell, May 21, 1920, Cornwell Papers, WVRHC.

9 "Flurry Caused By Invasion of Head of Mine Workers," *Bluefield Daily Telegraph*, February 1, 1920; The term "mine wars" generally refers to the episodic attempts to organize West Virginia's coal fields by the United Mine Workers of America. According to one of the men who popularized the phrase, H. B. Lee (who served as state attorney general, 1925–1933), there were four mine wars in West Virginia between 1900 and 1933: Lee, *Bloodletting;* Commonly used in feud lore, the phrase "stompin' ground" was immortalized in John Sayles' 1987 film, *Matewan* and a t-shirt sold by a Matewan merchant after the film's release.

10 "Twelve Men Killed in Pistol Battle in West Virginia," *New York Times*, May 20, 1920. The actual number of deaths was ten, although several more men were wounded, either during the battle or in related incidents nearby; Arthur Gleason, "Private Ownership of Public Officials," *Nation* 110 (May 29, 1920): 724–725.

11 *United Mine Workers Journal*, 31 (June 1, 1920): 5; Winthrop D. Lane, *Civil War in West Virginia: A Story of the Industrial Conflict in the Coal Mines* (New York: B. W. Huebsch, 1921 [reprint 1994]), 15; "Personal Views of Senator Kenyon," U.S. Senate, Committee on Education and Welfare, *West Virginia Coal Fields: Personal Views of Senator Kenyon and views of Senators Sterling, Phipps, and Warren . . .* (Washington: GPO, 1922) [67th Congress. 2d Session Senate Report 457]; Lane's pamphlet consisted of articles first published in the *New York Evening Post* in the spring of 1921. Senator William S. Kenyon of Iowa chaired the subcommittee that investigated the Massacre and strike; Source for "industrialism" title of essay: Arthur Warner, "West Virginia—Industrialism Gone Mad," *Nation* 113 (October 5, 1921): 372–373; Source for "serfdom": Neil Burkinshaw, "Labor's Valley Forge," *Nation* 111 (December 8, 1920): 639; Gleason, "Private Ownership," 724; Gleason, "Gunmen in West Virginia," *New Republic* 28 (September 21, 1921): 90–92.

12 Malcolm Ross, *The Death of a Yale Man* (New York: Farrar and Rinehart, 1939), 96. Ross also wrote one of the now classic accounts of the decline of the coal industry, *Machine Age in the Hills.*

13   A. F. Hinrichs, *The United Mine Workers of America and the Non-Union Coal Fields* (New York: n.p., 1923). This was apparently a printing of Hinrichs's dissertation for Columbia University's political science department.

14   Herbert Gutman, "Work, Culture, and Society in Industrializing America, 1815–1919," in Herbert Gutman, *Work, Culture and Society in Industrial America: Essays in American Working Class and Social History* (New York: Vintage Books, 1976 [1966]): 3–78, 53; Peter Friedlander's *The Emergence of a UAW Local* as quoted here is from a quotation in Paul Thompson's *Voice of the Past: Oral History*, third edition (New York: Oxford University Press, 2000), 293; Dwight B. Billings and Kathleen M. Blee, *The Road to Poverty: The Making of Wealth and Hardship in Appalachia* (Cambridge: Cambridge University Press, 2000), 14.

I

"BLEEDING MINGO": 1895–1911[1]

"All history is local."
– *Herbert Gutman*

ON OCTOBER 20, 1898, Henry Clay Ragland, the disgruntled editor of the *Logan Banner*, declared that Mingo County had been carved from Logan County three years earlier by a "Republican Statehouse conspiracy." The close association between the state's Republican Party and the rail and coal interests that developed southern West Virginia led most historians to accept Ragland's complaint for nearly 100 years. By the 1970s, a conceptual paradigm not uncommon in labor history had evolved, defining West Virginia's five southernmost coal-producing counties as a single non-union entity where miners slaved in the service of distant corporations that denied their workers basic constitutional rights, all in the name of mining coal as cheaply as possible. Because the early histories of McDowell, Logan, Fayette, Raleigh, and Mingo counties were believed to be interchangeable, place-specific root causes of events such as the Matewan Massacre were considered immaterial to understanding how and why they happened. Moreover, historians ignored the meaning and impact of Mingo's unique political, economic, and social development.[2]

Almost any Mingo County story needs to start with a setting of the political scene. From the time of its creation in 1895 until the resolution of the Williamson-Thacker strike in 1922, no single political party dominated Mingo. Within each party, rival factions consisted of mixed coalitions of natives and outsiders. Traditional mountain politics survived into the industrial era, but the clan patriarchs gave way, not to the coal operators, but to party bosses who derived their power not only from the patronage of the coal companies, but also from their ability to manipulate old style politics to keep challenges to the new order out of public discourse. As a result, the most powerful politicians in the county were masters of seemingly syncretistic sources of influence. For example, Democrat Wallace J. Williamson and Republican Greenway Hatfield were native elite politicians who actively and lucratively participated in Mingo's founding and development.[3]

Between 1895 and 1911, several hallmark elements of Mingo County politics emerged: the battle for political dominance between Williamson and Matewan; a propensity for election day violence, especially in Matewan; and the institutionalization of the corruption of the political process through diversionary subversions on election days and the outright purchase of Mingo's loyalty. Participation by industrial elites in Mingo politics did not substantively affect these systemic patterns of behavior. The syncretistic rebirth of the Hatfield family's political power in the county added a labyrinthine corruption that defied the corporate will; put more simply: more often than not, the coal elite supported the losing, not the triumphal faction.

Original Logan County
(Mingo Created 1895)

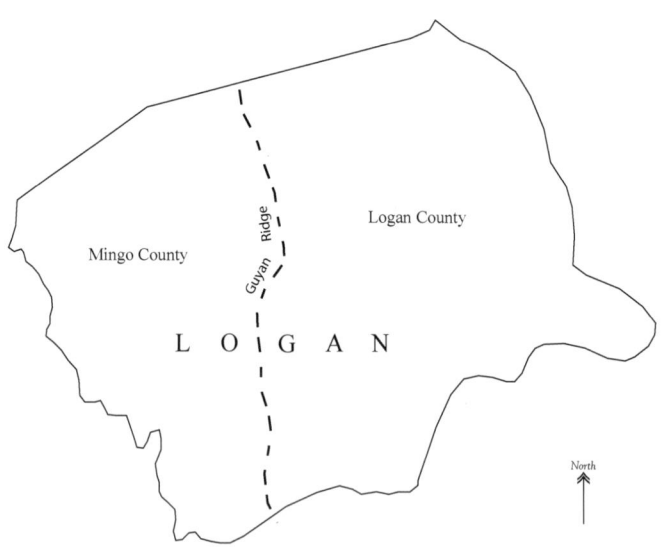

Logan County, West Virginia, ca. 1892. *The dashed line indicates where Mingo was carved from Logan in 1895. [Created by E-S Bonita Bonar]*

The last West Virginia county to be created, Mingo was carved from Logan County for a variety of political and economic reasons. Politically, the mid-1890s was a tumultuous period in West Virginia politics, and the bill creating Mingo was passed in the first legislative session dominated by the Republican Party in more than 25 years. For decades, historians have assumed that the ascendancy of the Republican Party, in both the state government and in southern West Virginia, precipitated the cleavage of the Democratic Party's most loyal county. However, a closer examination of the motivations and actions of the principal players in Mingo's early development reveals that both Republican and Democratic politicians, as well as industrial entrepreneurs, sought to manipulate the new county's founding to their advantage. Unlike in Logan and McDowell counties, political control of Mingo County shifted from one party to another with nearly every election. To know why political influence was so divided in Mingo County, the story of the county's founding has to be reexamined.[4]

While it cannot be disproved that Republicans were behind the division of Logan County, previously ignored sources reveal that several Democrats also played significant roles in the matter. In his memoirs, the man who proposed the bill that created Mingo County, Dr. Sidney B. Lawson, claimed that Democrats supervised the division because they feared that to not do so would "either make the new or leave the old county more or less of Republican complexion." According to Lawson, he and another Democrat manipulated census and survey records in an attempt to make two Democratic counties out of one. Lawson's summation of Mingo's creation as the result of "considerable conniving, many hours of tiresome work and a whole lot of friendly cooperation" reveals that Democrats may well have been the active force behind the creation of the new county. At the least, Lawson saw the possibility for Democratic gain. While a lifelong Democrat, Lawson was also a close personal friend and professional associate of Henry D. Hatfield, who became southern West Virginia's most famous native Republican. Lawson and Hatfield were later cofounders of the Hatfield-Lawson Hospital in Logan.[5]

Another cross-party friendship stood to benefit from creating a new county. Wallace J. Williamson, an emerging Democratic force in western Logan County, and state senator James A. Hughes of Huntington, the

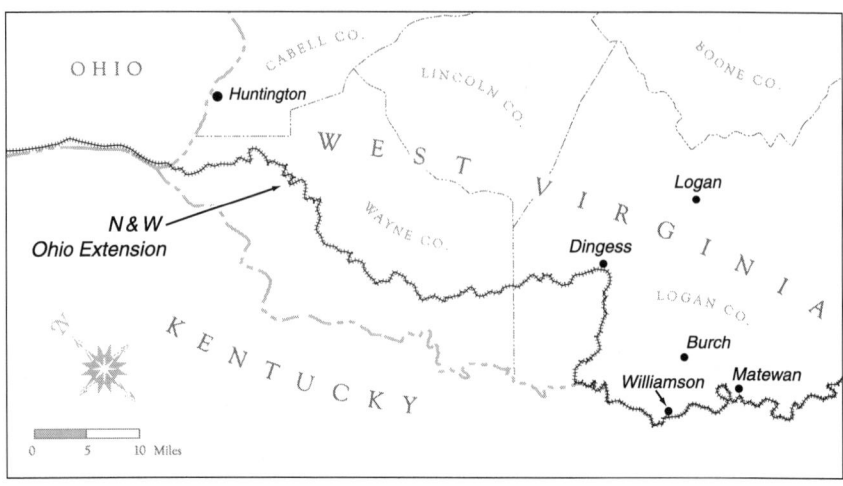

The N&W Ohio Extension. *The Ohio Extension of the Norfolk & Western Railroad.* *[Than Saffel/WVU Press]*

Republican brother-in-law of Williamson's closest business associate Z. T. Vinson, have also been credited with separating Mingo from Logan. By the early 1890s, the westward extension of the Norfolk & Western railroad (the N&W) ignited speculative activity throughout southwestern West Virginia. In 1891, Williamson and Vinson incorporated the first mining company of the new Thacker coalfield; soon thereafter, their consortium purchased the Williamson family farm and transformed it into a town. The Williamson Mining and Manufacturing Company also "paid all of the expenses for the surveys made by the engineers to make a separate county." When the N&W designated Williamson its western terminus, the new town was poised to eclipse the larger, older Logan Courthouse now remotely distant from the path of development. However, despite the influx of capital, including his own, and connection to powerful state leaders, Wallace Williamson failed to gain political control of Logan. The county remained under the control of a rival Democratic faction. Williamson's town, with its railroad connection, coal companies, bank, and other businesses, was a ready-made county seat, if only there could be a new county. The Republicans' capture of the statehouse afforded Williamson the opportunity for which he had been waiting for nearly four years.[6]

In January 1895, Dr. Sidney B. Lawson, whose family lands bordered the

Williamson farm on Pond Creek, took his seat as a first time member of the West Virginia House of Delegates and proposed the division of Logan County. The bill passed on January 23 and was approved by Democratic governor William Alexander MacCorkle on January 30, 1895. Williamson was named county seat "until otherwise provided by law," and three county commissioners were appointed to serve until the general election of 1896. For helping secure the passage of Lawson's bill, Williamson allegedly rewarded Hughes's "cooperation" by allowing him to name the new county's first sheriff. Despite political collusion being the source of Mingo's creation, politics in the new county rapidly descended into a fractiousness that lasted for decades.[7]

Between February 1895 and November 1896, when the first county elections were scheduled, a blueprint of Mingo's political future took shape, a pattern hallmarked by bitter rivalry between the county seat and outlying towns and districts, vicious factional infighting, and elections so corrupt that the nickname "Bleeding Mingo" appeared a decade before labor strife ever occurred.[8]

One of the primary reasons for the political turmoil that beset Mingo at its birth was the competition over the permanent location of the county seat. Although as the temporary county seat and N&W terminus Williamson had the upper hand, other communities fought for the designation. Because outsider capitalists entered the Tug Valley at roughly the same time from different directions in 1891, dual and dueling alliance systems emerged and thrived. Scarcely a month after the creation of the county, a public meeting to challenge Williamson's right to remain county seat was held in Burch. Before Williamson and Matewan were even founded in 1891, Burch had a population of 150, two general stores, a post office, a combination saw and grist mill, two carpenters, and a blacksmith. However, by 1895 both Williamson and Matewan had surpassed the older community in population and occupational diversity. Burch, also known as Rockhouse, was a tiny village off the rail line, "with only its central location to recommend it." In the years between Mingo's founding and the Massacre, Burch, like several other older communities in the county off the railroad or without coal development, stagnated as newer communities prospered and grew.[9]

As the only other significant contender for county seat in 1895, Matewan had less than half of Williamson's population and business development. However, because it was the political center of the Magnolia district, because it was the home base of the county's first prominent Republicans (including several Hatfields), and because the first large coal operation at nearby Thacker rivaled Williamson in size, Matewan had to be considered a serious rival. When the ballots were cast for the permanent county seat in 1896, Williamson received 821 votes, while Matewan received 804. Although a victory for Williamson, the near parity in the number of votes presaged the future temper of county politics. The balance of power in the county would swing back and forth between Williamson, the county seat and financial/industrial center, and Matewan, the center of the county's second largest industrialized district.[10]

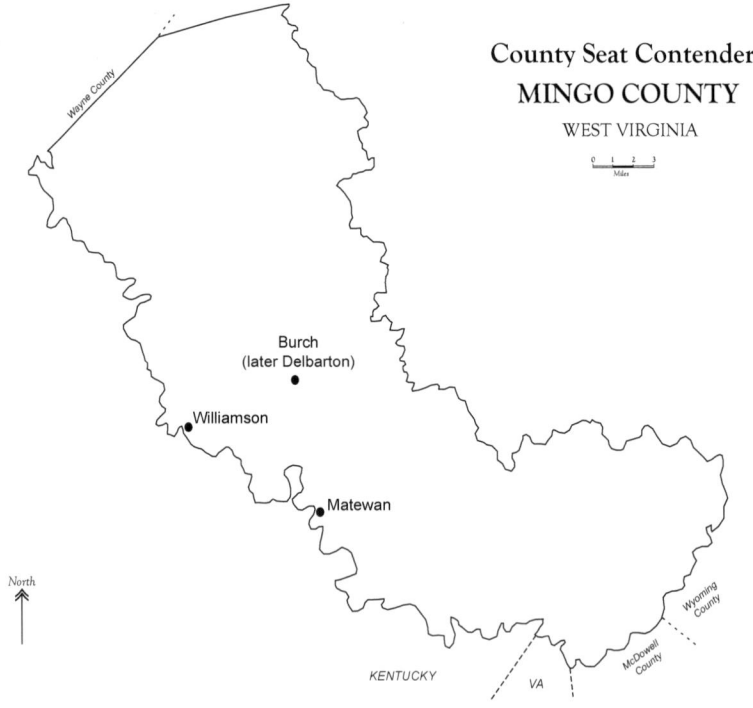

Mingo's county seat contenders, ca. 1895. *Three communities: Williamson, Burch, and Matewan contended for selection as the seat of government for the new county. [Created by E-S Bonita Bonar]*

A second reason for continuous upheaval in Mingo County politics between 1895 and 1920 was the factional infighting and party jumping engaged in by Republicans and Democrats. Although the founders of Williamson were Democrats, the other Democrats in what became Mingo County considered them interlopers. In the quarter century between the county's founding and the Massacre, Democrats from the "rural districts" of the county often coalesced into a minority opposition. The issues that fostered internal opposition were the distribution of patronage positions and the allocation of county and state funds for roads and similar infrastructure improvement projects. Much the same occurred within the Republican Party, except that factions drawn from the Williamson coal elite and disgruntled anti-Hatfield insurgents capitalized on the politically schismatic terminology of the era to present themselves as reformers.[11]

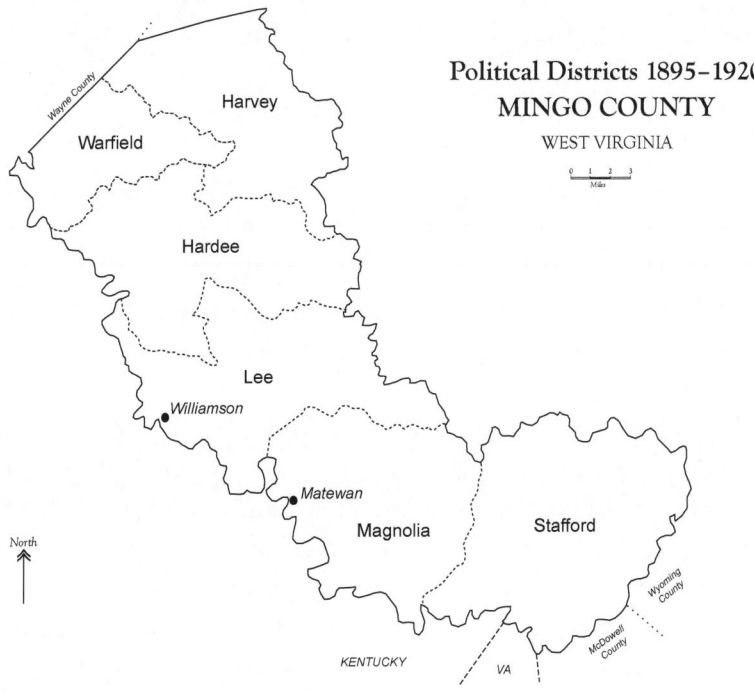

Mingo County's political districts, ca. 1895–1920. *Until the mid-teens, Williamson was part of Lee district; thereafter, the city was its own district. [Created by E-S Bonita Bonar]*

In the chaotic scramble for power in each election season, violence, sub-terfuge, and bribery became regular occurrences in Mingo County, most commonly in Matewan and the outlying communities associated with Matewan at the time. The county's competing factions also orchestrated diversionary conflicts in order to diminish their competitors' presence at important political events. The first illustration of this behavior appears in a report by a leading southern West Virginia Republican to then-Governor A. B. White. In his report, this is how Edgar P. Rucker described the 1902 Mingo County Republican convention:

> The Caldwell crowd, composed of a few respectable Republicans, a bunch of hirelings, and a crowd of Democrats, tried to create all the confusion possible in our Mingo Convention, but were out-generaled and outvoted. . . . The Caldwellites engineered the ap-pearance of a bolt to induce our people to compromise in order to avoid a split. [But] Scott is clearly entitled to the delegation from [Mingo] and will get it.[12]

The final noteworthy incident in the Republican campaign of 1902 underscores both the character of Mingo County politicians and how the state Republican leaders kept the local Republicans in line. In 1902, B. Randolph "Dick" Bias was a young coal company attorney serving as postmaster of Williamson when state party leader Nathan B. Scott named him the Republican candidate for the sixth district state senate seat. As described by his detractors, Bias, despite having a majority of 4,000 loyal Republican votes in the district, "was avaricious enough to demand of Senator Scott $3,000.00 to make his campaign." Scott replied that "he was not a national bank and that [Bias] had better get down and out if it would cost him that much to make his campaign." Scott's lieutenants soon found a replacement for Bias—W. H. H. Cook, a Baptist minister from Wyoming County. Cook won the state senate seat and proved to be an ideal party hack. According to his critics, Cook rarely attended legisla-tive sessions, and only appeared to vote for Scott.[13]

The factions that dominated the Republican and Democratic parties in Mingo County until 1920 were well established by 1904. The Republicans were divided between the "Old Liners" and the "Regulars" (see Table 1), while the Democrats were split between the courthouse elite, referred to as the "City Ring" and the leaders of smaller communities, who were called the "County Democrats." All four factions included native and outsider elites, all of whom in one way or another were dependent on Mingo's emergent industrial economy.[14]

"Old Liners"	"Regulars"
Hiram S. White	Greenway Hatfield
Everett Leftwich	M. Z. "Mont" White
B. Randolph Bias	A. B. Hatfield
S. T. Lambert	O. H. Booten
J. R. Booth	J. K. "Jack" Anderson
Harry Scherr	R. W. "Bob" Buskirk
	Bob Simpkins

Table 1. The Republican Factions of Mingo County, 1908–1920
Sources: "Condensed Facts," "The Old Liner," "Williamson Daily News" [WVRHC]

Democrat Wallace J. Williamson and Republican Greenway Hatfield, as descendants of pioneering Tug Valley residents, were true natives who also directly affected the county's development—Williamson as a returning capitalist entrepreneur and Hatfield as a land agent for an outside corporation. Their factional opponents were G. T. Blankenship and Hiram S. White, respectively, both of whom were "quasi-natives." An N&W railway agent, Blankenship came to Mingo in 1900 and married into the Chambers family of Matewan, who themselves had only arrived in the Tug Valley in 1870, but had intermarried with the Hatfields. The brother of West Virginia state geologist Dr. Israel C. White, Hiram S. White had settled in Logan Courthouse in the 1870s and then moved to Matewan in the early 1890s. Hatfield and Williamson built "machines" whose domination of local politics was periodically challenged by the factions led by White and Blankenship. White's "Old Liner" faction consisted mainly of disaffected Republican coal industry elites. The bolts of White's faction also followed the pattern of the schismatic national Republican politics

State Senator Hiram S. White. *Brother of West Virginia's State Geologist I. C. White, H. S. became one of the leaders of the anti-Hatfield "Old Liner" faction in Mingo County. [WVRHC]*

between 1900 and 1912. By contrast, Blankenship and his Chambers in-laws gradually established a reputation as the advocates of the people most threatened by the coal industry: the miners, farmers, and independent small businessmen of the county.[15]

Mingo County's first well-documented election dispute involved the general election of 1904. The elections of the sheriff, house of delegates representative, county assessor, prosecuting attorney, county commissioner for the short term, and county commissioner for the full term were all contested. Determining which party would control the county took more than 17 months. A total of seven cases detailing the conduct of the election went before the West Virginia State Supreme Court of Appeals. Although there were several other reasons why the Democrats contested the election results, including the illegal creation of precincts and tampered ballots, the decisions of the high court focused on the 25 irregularities that occurred in Matewan.[16]

Two examples capture both the atmosphere at the poll in Matewan on November 8, 1904, and the Supreme Court's disregard for the evidence of fraud. The court dismissed as immaterial the accusation that R. W. Buskirk "usurped" the legitimate poll clerk by "jabbing his pistol down in his pocket" and proclaiming, "By God, if I don't be poll clerk, nobody else will be!" The court also disregarded the testimony of one of the morning election commissioners who claimed that he failed to vote because when he went to vote, he was first told he could not vote, and then told the denial was only a joke. Matewan's local lore is full of the characterization, "he was the kind of fellow who would be laughing and joking one minute and would kill somebody in the next." The Hatfields in particular were said to have this type of temperament and often engaged in intimidation laced with humor and the threat of violence.[17]

By rejecting only a portion of the suspicious ballots from Matewan, the judges' decision threw the entire election from Democrat H. H. "Hi" Williamson to Republican E. E. Musick, even though if the Matewan results had been dismissed completely, Williamson would have won. The majority of the court's justices decided not to invalidate the entire election at Matewan because they feared that in so doing they would be "disfranchising [sic]" the voters of Matewan. However, both Chief Justice Poffenbarger and Judge McWhorter dissented angrily from the Court's decision. Judge Poffenbarger chastised his fellow judges for their decision and cited as proof of their error the following assertion from the majority's own decision: "When it becomes apparent that an election is a subversion rather than expression of the will of the people, or that the result is attended with such uncertainty that it may not be ascertained, the election should be set aside."[18]

By not holding the politicians of Mingo County accountable for their actions, the West Virginia Supreme Court of Appeals not only validated acts of subversion, but also ensured the future corruption of legitimate political processes in Mingo County. The Republicans had been allowed to regain control of the county for the first time in a decade through legitimized theft. The lasting effect of the 1904 election dispute was the creation of a blueprint for how to manipulate Matewan's elections and thus influence the county election as well.[19]

After the state legislature created Mingo County, coal men initially figured prominently in local politics. J. K. Anderson, the general manager of the United Thacker Coal Company holdings in southern West Virginia, was one of Mingo's first three county commissioners. S. T. Lambert, who worked as a superintendent for several mines between 1893 and 1918, served as Magnolia District's first justice of the peace. Other coal men, such as James Little, represented their districts in the county organization. Little disappeared from the political records after his mine closed. Lambert and Booth, by aligning with a losing faction, asserted little power during the reign of their factional opponents. However, J. K. Anderson, who represented one of the largest coal interests in the county, backed the Hatfield faction and influenced local politics through them.[20]

In 1908, a statewide schism in the Republican Party highlighted the growing division in Mingo County's own Republican ranks and the eclipse of the anti-Hatfield coal men. Although the first hint at the cleavage had come during the first county elections in 1896 when Hiram S. White bolted to support the Democratic candidate for sheriff, the deep-seated reasons for the antipathy between the "Old Liner" and "Regular" factions had grown over the course of the intervening ten years. In the interim, control of the party passed from the "Old Liners" and their coal-elite allies to the "Regular-Renegades," a group that eventually coalesced around Greenway Hatfield, a native elite with his own industrial ties. The events that marked the transference of power from the Old Liners to the Regulars fell largely between 1906 and 1908. For nearly a decade after the 1908 election, the various members of the deposed Old Liner faction alternated between acquiescence to Hatfield rule and co-option of schismatic national Republican third party issues in their efforts to overthrow the Hatfield-dominated Regular faction.[21]

Despite bolting the party in 1896, Hiram S. White's Old Liner faction dominated the county Republican Executive and Central Committees for several years. Like White, the other Old Liners had been drawn to Mingo because of its industrial development. The Old Liners based in Williamson were represented by B. Randolph Bias and Everett Leftwich, attorneys who represented and invested in the local coal companies and brokered land development deals. The other coal industry men were entrepreneurs such as James Little or corporate managers like S. T. Lambert. The issues championed by the Old Liners reflected turn-of-the-century business progressivism: fiscal accountability in local government, opposition to blatant patronage system abuses, and equitable expenditure of tax revenue. The Old Liners had to vie for control of Mingo not only with the Democrats who controlled Williamson (and the county until 1904), they also had to fend off insurgents from within their own party.[22]

From 1900 until 1906, every addition of an Old Liner to an executive position in the county party was matched by one from the Regular faction. Names of the Regulars reveal the primary source of their influence: S. A. Ferrell, A. G. Rutherford, and John A. Sheppard. Ferrell, Rutherford, and Greenway Hatfield, who joined them on the Executive Committee in 1906, were scions of pioneering Tug Valley families. Sheppard, although a Virginia native, had gained local prominence and trust during the early phase of the infamous King Land Case. However, the difference between the Old Liners and the Regulars cannot be distilled down to fights for primacy between simplistically and diametrically opposed industrialists and traditionalists, natives and outsiders, reformers against machine men. Moreover, examples from their 1908–1910 factional warfare illustrate that neither group's leaders allowed principle to interfere with the pursuit of power.[23]

The primary excuse for the open confrontations between Mingo's Republicans in 1908 centered on the selection of the 1908 Republican candidate for governor. Of the two leading candidates, Secretary of State Charles W. Swisher and Attorney General Arnold C. Scherr, Scherr had a direct connection to Mingo. His son Harry, after graduating from West Virginia University's law school, had moved directly to Mingo in 1905 and rapidly ascended the party ranks. The younger Scherr, who resided in Williamson, had aligned with the Old Liner faction led by Hiram S. White. The fight over which candidate Mingo's Republicans would support led to the exposure of the other reasons for the division between the Old Liner and Regular factions. The local issues that disrupted county-level politics included accusations of fiscal corruption and the growing dependence of one faction on African-American votes.[24]

The acrimony between the two groups deepened to the point that White's faction produced its own newspaper, *The Old Liner,* in Matewan, in order to excoriate their rivals for using elected office to enrich themselves and their own districts. For example, *The Old Liner* claimed that the poor people of Mingo County should oppose the road tax because all the money went to Guy White and R. W. Buskirk for roads between Matewan and Thacker, leaving "the rest of the county with poor roads and less development." *The Old Liner* also published the names of 12 Regulars who were bleeding the public coffers dry with "reimbursement claims." Among the alleged leeches were Mingo's incumbent sheriff E. E. Musick and Greenway Hatfield.[25]

When *The Old Liner* published "To The Honest Voters of Mingo County," it exposed the methods employed by the Regulars to manipulate election results. According to the circular, the Regulars used "election day rumors," "stories," and "affidavits" to distract voters, "hired ruffians . . . to bluff and bulldoze the decent," and ballot box stuffing to steal the election. The circular also urged citizens to arrest illegal voters and send them to Williamson.[26]

Despite the efforts of *The Old Liner,* the Regulars overwhelmed the Old Liner Republicans at the county convention. Their success stemmed from two significant developments. First, M. Z. White and Greenway Hatfield had masterminded a Democratic bolt that swelled the Regulars' ranks by 50. Second, the Old Liners were betrayed from within by a "Judas . . . who . . . sold out his party . . . to Mont and Jack," when Bias broke from the Old Liners to support the Regulars. Thus, at the 1908 county convention, the Regulars orchestrated the selection of M. Z. White as county chairman and the deposition of both Harry Scherr and B. Randolph Bias from Mingo's Republican Executive Committee. White's victory fulfilled *The Old Liner*'s fearful prophecy that "a complete change in the control of affairs in the county" would come if the Regulars triumphed. From that time until the 1930s, either M. Z. White, Greenway Hatfield, or another member of their faction served as the head of Mingo County's Republican Party. Only the coal men who made peace with the faction's methods of achieving and maintaining power substantively affected Republican politics in the county.[27]

In 1908, the only exceptions to the factional fighting in Mingo were the personal advertisements of two coal industry candidates for local office. Neither man attacked his competitor nor made overt reference to his own factional association. However, each man revealed his political alignment by the elements of personal biography stressed in the advertisements, and these revelations illuminated where both stood on the current conflict within the party.

J. R. Booth and S. T. Lambert were pioneer coal men in Mingo County. Booth had been the superintendent of the Pearl Mining Company at Dingess in the northeastern section of the county, while Lambert had been superintendent of several mines at or near Thacker in the south-central section of the county. Both were also longtime members of the Republican Party. In 1908, Booth, who was no longer working as a super-intendent, ran for the House of Delegates. Lambert, the superintendent of the Mate Creek Coal Company, ran for sheriff. Booth defined himself as a self-made man who had worked as a coal miner, foreman, and mine su-perintendent and now worked on his own as a lumberman. The message of his advertisement to the voters of Mingo was, "I am a common man like you." Lambert, like Booth, had come to the county in the early 1890s because of opportunities in the coal industry. However, he did not make any overt reference to coal. Lambert's advertisement stressed that he had come to Thacker from Kentucky at the urging of Republican friends and had been a loyal party man ever since. Thus, without emphatic asser-tion, both men revealed their factional loyalty. Both Booth and Lambert aligned with Hiram S. White's Old Liner faction which claimed to repre-sent the legitimate party.[28]

The 1908 campaigns of Booth and Lambert inadvertently reveal why the "coal men" in Mingo County did not wield the local political power enjoyed by the industrial elites in the surrounding counties. Mingo's coal industry elite was a diverse group divided by nativity, religion, education, party affiliation, and occupational status. Moreover, whether entrepre-neurs or corporate managers, like Booth and Lambert, the county's coal men also disagreed about which faction to support.[29]

One man encapsulated the root cause of the dilemma faced by the coal elite. Henry Drury "Drewy" Hatfield, who had been identified as the "promising pup" of the family litter as one family descendant recalled, was a nephew of Devil Anse and the family spared no expense in cultivating his abilities. Something of a child prodigy, Drewy departed the Tug Valley for college at age 15. Liberally educated at several colleges and universities, he returned to the Tug Valley in 1894, a doctor at age 19. Hatfield began his medical career in the southern West Virginia coalfields as a company doctor, and his lifelong commitment to quality healthcare for miners and their families is evidenced by the miners' hospitals he helped establish. However, Henry D. or "Doc" Hatfield, as he was known outside the Tug Valley, also entered politics. It was in this public sphere that the twin influences of his family background and professional training at times warred and at other times dovetailed to mold him into both one of West Virginia's most successful reformers and one of its most corrupt politicians. Celebrated by southern West Virginia's coal men as the shining example of how industry could positively transform the future of mountain people through educational and professional opportunity, Henry D. Hatfield also confounded his patrons by refus-

Henry Drury Hatfield. *Nephew of "Devil" Anse Hatfield and West Virginia governor (1913–1917), heralded for decades as a prominent Progressive physician and politician. "Drewy," as he was called by the family, was one of the state's most successful machine politicians. [WVRHC]*

ing to abandon the corrupt politicking of his extended family.[30]

The sheer size of the Hatfield family made it a force to reckon with in southern West Virginia politics. By the time Henry D. Hatfield first ran for state office in 1910, the Hatfields and/or their relatives were a dominant force in the politics of three counties: Mingo, Logan, and McDowell. Through his mother, Betty Chafin, Henry D. was related to the Democratic Chafins who ran Logan County. Joe D. and Tennyson or "Tennis," two of Devil Anse's sons and thus double first cousins to Henry D. Hatfield, were part of the Logan County Chafin machine until the mid-1920s. Henry D.'s second cousins, brothers William "Bill," and McGinnis "Mac" Hatfield, were an emerging force in McDowell County Republican politics. Henry D. himself had helped found the Republican Party in Mingo in 1895–1896, before moving to McDowell County. Whatever the reason for his departure, once Henry D. Hatfield moved to McDowell, he and cousins "Bill" and "Mac" Hatfield forged an alliance with the coal and industry elite of that county. Henry D.'s older brother Greenway Hatfield and their brother-in-law M. Z. White led one of two major Republican factions in Mingo.[31]

Eventually the twin pillars of non-union southern West Virginia, Logan and McDowell counties, while as politically corrupt as Mingo, were far more quiescent in the new industrial order. When the coal industry entered Logan in 1904, it was considered an efficient expedient to pay the Chafins (of whom Don Chafin was the emerging leader) to maintain order, which they did for the next 30 years. Thus Logan politically (and at least nominally) remained a Democratic county. Conversely, the Republican primacy in McDowell depended on the loyalty of the county's African-American population, who were imported by Republican coal entrepreneurs and whose percentage of the total population exceeded the combined figure for native and foreign-born whites. In McDowell's coal camps and predominantly black towns like Keystone and Kimball, the Hatfields helped build a machine that provided patronage and protection to keep the African Americans in line for the next several decades.[32]

By 1910, when Henry D. decided to run for a seat in the state senate, his older brother Greenway had carved his own niche among the Republican elite of their home county. The abuses orchestrated by the Hatfield brothers to win Mingo and McDowell County in the 1910 election were so flagrant that Rankin Wiley, the defeated candidate for the Fifth District United States congressional seat, sought and secured a congressional investigation of the election proceedings in both counties. The investigation's report reveals that while "Doc" Hatfield espoused support for a variety of reforms, including the institution of primaries, workers' compensation, and mine guard legislation, he also used the tools of a corrupt party boss to enhance his own power.[33]

In McDowell County, "Doc" Hatfield had refused to appoint any

Democratic election supervisors. The congressional committee also noted concern over Wiley's inability to campaign in McDowell because he had been threatened with bodily harm. However, the committee seemed more interested in the charges leveled against the Hatfields' behavior in Mingo County. After noting that nine precincts were troubled on election day, the activities in Matewan drew the most attention. The precinct was so notorious, the committee report observed that people went to extraordinary lengths to secure a fair election:

> Conditions had in recent years become so outrageous at this precinct that a number of women belonging to the Woman's Christian Temperance Union determined to put a stop to fraudulent elections there, and with the finest spirit of American womanhood, repaired to the polling place at daybreak and begged the election officials to see to it that a fair election was held that day.

Despite the women's efforts, the election at Matewan proceeded as corruptly as ever. African-American men known as "floaters" were brought to the town in advance of the election and kept there several days. On election day they were marched to the poll, and when they forgot the name of the registered voter they were replacing, Greenway Hatfield shouted it out for them. To keep track of how many times the men voted, the Democratic election commissioner marked each man on the back with a piece of chalk; at the end of the day one man had 12 marks. When a coal company doctor protested voting the "darkies" this way, "he was struck in the face and driven from the polls." Of the 327 votes cast at Matewan, 237 were cast by African Americans, even though Matewan had only ten African Americans registered to vote at the time.[34]

The investigating committee was most shocked to learn that the West Virginia legislature, instead of censuring "Doc" Hatfield, elected him president of the state senate. However, despite their own disgust with Hatfield's behavior, the congressional committee declared that the lack of bipartisan oversight of the McDowell polling stations did not in and of itself constitute fraud nor prove that fraud had been committed. The committee did award the plurality in Mingo to Wiley. However, this was not enough to turn the entire district's vote to Wiley's favor. Despite openly acknowledged irregularities, and in some cases proven fraud, Wiley's opponent James A. Hughes retained his seat as United States congressman, thanks to the Hatfield machine of Mingo and McDowell Counties.[35]

An examination of one bill passed during the 1911 legislative session exposes the secret of Hatfield's success. The West Virginia Primary Election Law transferred the responsibility or the right of picking a party's slate of candidates from the county and state conventions to a direct primary. Like similar legislation around the country, the West Virginia bill was meant to "democratize" the candidate selection process and thereby undermine the influence of "cliques" and machines. However, the bill's limited ability to fight corruption was noted almost immediately after its passage. One month after the end of the 1911 session, John J. Cornwell, former Democratic gubernatorial candidate and editor of the *Hampshire Review*, wrote an editorial about the law and its impact. Cornwell prophesied that "party primaries would usher in a period of unparalleled disunity and corruption in Republican politics." Subsequent primary elections in Mingo County affirmed the accuracy of Cornwell's prediction.[36]

Experienced in the art of corruption, the Hatfield brothers, Henry D. and Greenway, would rise to the very pinnacle of West Virginia politics by 1912. Their long-term critic and political nemesis, John J. Cornwell, would draw on the anger unleashed by the Hatfields' unscrupulous behavior to unseat them in 1916. As the only Democrat to serve as West Virginia governor between 1896 and 1932, Cornwell would be at the helm when "hell broke loose" in Mingo in 1920.[37]

One of the most profound influences on Mingo's political marginalization was the unfulfilled promise represented by its coalfield. Despite being the first West Virginia coalfield to open mechanized, the Thacker coalfield never placed among the state's top coal producers. Developed during a period of profound change in the American coal industry, the precise timing of the Thacker coalfield's opening and development set it apart from its larger and more important neighbors. Long overlooked by historians, these differences not only affected political and social relations in Mingo County; they also influenced future industrial relations.[38]

Like the surrounding coalfields, the Thacker coalfield owed its founding to the expansion of the railroad. Also like the other southern West Virginia coalfields, investment shared the same corporate parentage, and the men who directed the coalfield's development subscribed to the same anti-union philosophy. However, the story of the Thacker coalfield's development diverges from the generally acknowledged pattern of southern West Virginia's coal history. Unlike the older smokeless coalfields to its east, the Thacker coalfield did not enjoy paternalistic support from the Norfolk & Western in its early stage of growth and expansion. Also, the coal produced in the Thacker coalfield, although targeted for the same markets as smokeless coal, was of lower quality, and thus less competitive. In contrast to the younger Guyan coalfield to its northeast, the Thacker coalfield was utterly dependent on the Norfolk & Western to export its coal. But most important, from the outset, less powerful men guided the Thacker coalfield's development. The interests of the large corporations' mines were represented by managers, while none of the coalfield's pio-

Mingo County, West Virginia, ca 1920. *Shown here are the communities most affected by the 1920–1922 strike. [Created by E-S Bonita Bonar]*

neering entrepreneurial operators ever acquired the position or influence of their contemporaries in the surrounding coalfields.

Cumulatively, the timing of the Thacker coalfield's opening and the differences between it and the other coalfields in southern West Virginia created a siege mentality among the men who controlled mining in Mingo County. As producers of a less competitive coal, the companies of the Thacker coalfield had to fight perennially for survival, and their adversaries included not only other coalfields but also an indifferent railroad. Lacking decisive, on-site leadership, the coal elite of Mingo County lived under the gun, striving to meet the production expectations of far removed corporate interests, or reacting to the decisions made by the coal men of the larger, better capitalized coalfields around them. Most prosperous during times of crisis in other coalfields, namely during strikes or union organization drives, the Thacker coalfield became increasingly dependent on maintaining its own non-union status.[39]

The same year that the mines in what would become Mingo County went into production, one of the nation's worst depressions gripped the United States. Throughout the coal industry companies responded to the lingering economic crisis by increasing production and imposing cost-saving measures, including wage reductions. As the crisis lasted into the closing years of the century and their employers showed no signs of restoring old pay rates, miners gradually abandoned their traditional policy of cooperation and began flocking to the fledgling United Mine Workers of America.[40]

In this atmosphere, the pattern of industrial relations in Mingo County's Thacker coalfield took shape. The long, slow recovery from the national depression also coincided with a broad-reaching transformation of the American coal industry. The growth and expansion of the coal industry in Mingo County not only obscured the differences between it and surrounding coalfields, but also disguised the long-standing systemic problems that precipitated the descent into crisis in the spring of 1920.

The economic ramifications of the Panic of 1893 profoundly affected the initial stage of Mingo County's industrial development, even though the Thacker coalfield continued to expand. Coal production more than doubled, but growth in the Thacker coalfield did not necessarily signify fiscal soundness for the coal companies, the Guyandot Coal Land Association, or the Norfolk & Western. Rather than cooperatively confront the impact of overproduction and overcompetition, coal men rapaciously sought ways to produce cheaper coal. As miners in the older coalfields came together and founded an organization to protect themselves from their employers' efforts to lower wages, many operators searched for areas where they could start fresh and avoid the issue.[41]

In this climate of crisis, the pattern of industrial relations in the Thacker coalfield was established. From the beginning, coal companies in Mingo County operated under a siege mentality, and deprived of the benefits enjoyed by the smokeless coalfields to the east, the operators of the Thacker coalfield were slow to organize in pursuit of common interests. The story of the United Mine Workers of America's early organizational efforts underscores not only the vulnerability of the Thacker coalfield but also the roots of the union's own tenuous relationship with Mingo's miners.

When the Norfolk & Western opened the coalfields of southeastern West Virginia, it required all of the companies served by its line to sell their coal through the N&W's own sales agency. While most of the benefits of this arrangement accrued to the railroad, the companies were compensated by other aspects of the deal. First, because their coal was sold under a single label, the smokeless operators benefitted from the rapid advance in demand for their product. Second, and most important, the N&W guaranteed the companies a minimum price for their coal. After the Panic, however, the continued instability of the coal market frequently required the N&W to undercut itself by compensating the coal companies for the shortfall between the prearranged price and the actual market value of the coal. After entering receivership in 1895, the N&W embarked on a new coal sales policy. In late February, the railroad abolished its coal sales company, the Pocahontas Coal Company, and turned over its coal sales contract to Castner, Curran, and Bullitt of Philadelphia, which had been the N&W's seaboard agent. The new arrangement in coal sales promised to help stabilize the railroad's operating costs, and raise its coal hauling profit margin.[42]

For the coal companies in the Pocahontas and Thacker coalfields, the withdrawal of guaranteed price supports removed an important safety

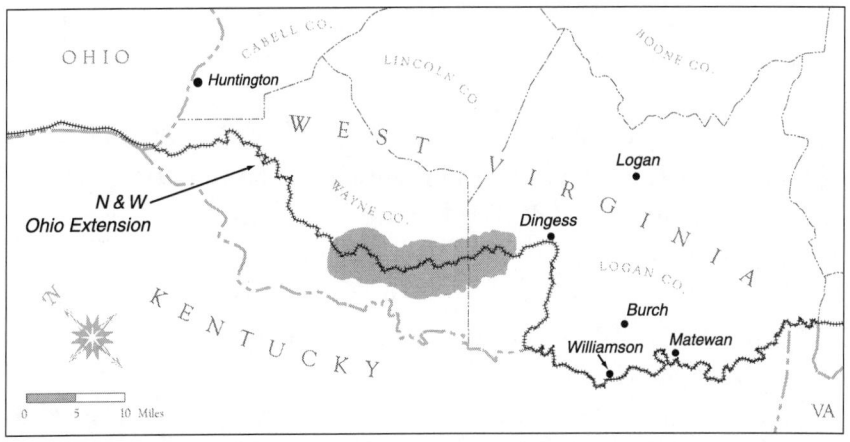

The Ohio Extension of the Norfolk & Western Railroad and The Guyandot Coal Land Association, ca. 1892. *The dark line indicates the rail line and the shaded area is roughly analogous to the holdings of the Guyandot Coal Land Association. [Than Saffel/WVU Press]*

net. In the midst of a failing market, the operators faced a further drop in coal prices. All of the N&W's southern West Virginia coalfields suffered from the loss of the guaranteed price support. The Thacker coalfield, only two years into production and lacking the established reputation enjoyed by the smokeless coalfields, was hit hardest by the N&W's decision. Forced into open competition much earlier in its productive life, the Thacker coalfield never escaped the shadow of the better-known and established smokeless coalfields.[43]

In 1899, the noted self-taught mining engineer and miners' advocate Andrew Roy published an account of a visit to the Thacker coalfield. Primarily upbeat, his article also analyzed the various disadvantages of the young coalfield. Chief among the disadvantages Roy singled out was the challenging relation between the operators and the railroad. Unable to forecast the outcome of such issues, Roy chose to highlight the "class-bending" engaged in by typical employees of the Thacker Coal & Coke Company:

> The miner keeps his dress suit in his own closet while at work in the mine. After bathing in the evening he dons this suit and

places his mining garb in the closet and emerges from the bath-
room as "clean as a new pin" and looking more like a congressman [than a miner].

The implications of Roy's depiction are clear. The miners of Mingo's
Thacker coalfield enjoyed the fruits of the latest innovations in American
mining. After working in one of Mingo's thoroughly modern mines,
Thacker's miners emerged from the company washhouse cleansed of the
coal grime that would otherwise mark their occupational status. Thus
liberated, the miners could move among local society, the equals of any
man. What Roy's observation does not reveal, however, is that his visit
coincided with a tumultuous time in the Thacker coalfield, when local
miners undertook their first efforts to join the United Mine Workers
of America.[44]

Cut adrift by the N&W, forced into competition with bigger, better-

Figure 1.4 Thacker Coal & Coke company town, ca. 1908. *[Payne, The Illustrated
Monthly West Virginian, 1908] [WVRHC]*

known coalfields, and denied the opportunity to emulate the sales strategy of these same competitors, the Thacker coalfield struggled to survive
the aftereffects of the Panic. Not unlike coal operators around the nation,
Thacker companies competed for skilled workers to help boost production, only to impose wage reductions as soon as they thought they could.
Thus, less than four years after opening, conflict between the miners and
the companies wracked the Thacker coalfield.[45]

Although the UMWA included West Virginia in its first organization
drive in 1894, the production crisis precipitated by the union's success in
Ohio, Illinois, and Indiana only fueled the growth of newer coalfields like
the Thacker coalfield. The miners' response to the change in the N&W's
coal sales policy in 1895 hints at a possible reason for their refusal to join
the 1894 strike. According to the *Bluefield Daily Telegraph*, the miners
of southern West Virginia, in recognition of the hardship faced by their
employers, offered to "concede to some wage reduction" in return for a
guarantee of full-time work. The cooperative stance of the Thacker miners
revealed their adherence to a traditional philosophy of labor-management
relations. Until the late nineteenth century, miners believed in a "har-
mony of interest," the idea that they and their employers had a common
interest in maintaining production. However, persistent adverse economic
conditions and evolving managerial philosophies negated the importance
of the miners' sacrifices. As a result, between 1897 and 1902, miners in
the Williamson-Thacker coalfield answered the United Mine Workers of
America's strike call six times.[46]

Scanty records limit analysis of the 1897, 1898, 1899, and 1900 strikes
in Mingo County. Brief in length, Mingo's 1897–1900 strikes reflect the
painful transition of turn-of-the-century American mining. The reasons
for these early strikes included demand for higher wages (a perennial
cause), the firing of a local union president, and safer work conditions.
While the 1897, 1899, and 1900 strikes ended in capitulation, the 1898
walkout ended when the operators hired replacement workers. The small
amount of available information on these local attempts to participate
in larger efforts seems to indicate that the 1897–1900 strikes in Mingo
County were plagued by a lack of cohesion and cooperation among the
miners. The operators' success in forestalling the unionization of their

employees stemmed not from grand strategy but rather from their ability to "starve" their men out or find area natives who were more than willing to hire on as replacements.[47]

Strike activity in Mingo's Thacker coalfield was transformed in 1901 when, as part of the 1900–1902 UMWA's effort to organize West Virginia, the union organized two subdistricts in southern West Virginia, one in the Thacker coalfield and another on the New River. Although these subdistricts rapidly amassed a combined membership of 5,000 miners, the strikes undertaken in 1901 and 1902 also collapsed in failure. Accounts of these strikes suggest that, just as later during the mine wars, the operators defeated the miners through "injunctions, deputies, strikebreakers, and intimidation." The now classic understanding of the unionization of southern West Virginia's miners also accepts without question the UMWA's condemnation of the miners for the failed strikes. However, the local story of these strikes exposes heretofore unacknowledged connections between the early organization efforts and the 1920–1922 initiative, among these, the partial culpability of the UMWA for their failure. The parallels between the two eras include the participation of the same individuals in the early and later strikes, the support of Mingo's sheriff for the strikers, and most important, the significance union organizers placed on the area's connection to the Hatfield-McCoy feud.[48]

The early success of the 1901 organization drive in the Thacker coalfield resulted from the miners' ability to attract support that undermined the strike-breaking efforts of their employers, at least initially. First, the miners persuaded imported replacement workers to join them or return home. Second, Mingo's sheriff refused to interfere in the conflict. Third, when the operators sued the union for unlawful enticement, the local court ruled against them. Despite these startling accomplishments, the operators prevailed, and the 1901 Thacker strike collapsed within months. Union organizers and officials condemned the miners for the failure, claiming that the miners "lost interest" in the union when they "discovered that they had to pay initiation fees." However, it was the president of the miners' local at Matewan, S. S. Morrison, who exposed the most critical obstacle to organizing the Thacker coalfield when he wrote to UMWA president John Mitchell: "We are not being treated right here . . . if there is not more attention paid to the actual needs of our men in justice to myself and men I will resign and in this declaration I voice the sentiments of all the officers of our LU [local union]." The UMWA's effort in the Thacker coalfield collapsed, not only because the men returned to work as non-union employees, but also because the national union failed to provide adequate funding and capable leadership.[49]

The operators tied the union up in the courts and used deputies to physically intimidate the organizers and miners in the other southern West Virginia counties. The organizers and the national union retreated before the operators' stiffening resolve led to pitched battle. As Matewan's local union president Morrison observed to President Mitchell, "Had we this company to fight only it would have been settled long ago, but right here in this Thacker coalfield we are fighting nearly every operator in the State of West Virginia." According to union organizer W. H. Crawford, the Thacker miners had told him that when they saw that to "fight was the only chance . . . they would take their Winchesters and defend themselves." To underscore the sincerity of his report, Crawford reminded his superiors:

> Thacker is in a region where they certainly will use guns to settle differences of opinion being the centers of the Hatfield-McCoy feud. Those people are there yet and both sides to the controversy are with the miners.

Weakened by dysentery and unnerved by the death of a fellow organizer (which he initially ascribed to poison), Crawford turned on the miners of southern West Virginia. Derided by the UMWA as violent but weak, the miners themselves thereafter became the sole scapegoats of the failed early efforts to organize southern West Virginia.[50]

Defeated in 1901, castigated by their own leaders, and still fighting their employers in the court system, Mingo's miners tried to unionize the Thacker coalfield in 1902 for the fifth time in six years. On June 7, 1902, 132 miners at four mines responded to a strike call from the UMWA. The number of men thrown out of work by the strike grew by 60 percent when the Grapevine mine locked out its 80 miners. Since 80 percent of the rest of the state's miners also responded, the UMWA focused its attention on the previously unorganized Fairmont coalfield of north-central West Virginia and the other southern coalfields. Left to their own devices, the miners of the Thacker coalfield for a time succeeded without assistance. Nearly 20 years later, the Williamson-Thacker operators' association acknowledged that in 1902, the Williamson-Thacker field had been organized "without much objection." But, isolated and unsupported, the miners lost their battle again in just over three months. On September 1, 1902, four days before it ended in the other southern West Virginia coalfields, the strike in Mingo was called off and the men returned to work still without union recognition or a union contract. Although the miners in Mingo County did not strike for the union again until July 1920, memories were long; Sam Artis, one of the union co-defendants in the 1901 lawsuit, reemerged a leader in 1920. The father-in-law of deputy sheriff "Toney" Webb, Artis was shot at Red Jacket on May 19, by the superintendent William Cummins.[51]

Although freed from the threat of unionization, the Thacker coalfield faced new problems shortly after the dawn of the new century. Still in the shadow of the older, larger coalfields to the east, the Thacker coalfield suffered additional pressure from the opening of the Guyan coalfield in Logan County. In less than a decade, coal mines in Mingo's parent county caught up with and surpassed those in the Thacker coalfield. The close of the coalfield's second decade found it caught between two large and powerful neighbors, the Guyan coalfield and the Pocahontas coalfield of McDowell County.[52]

Between 1880 and 1910, Mingo, Logan, and McDowell counties underwent a parallel experience in terms of industrial development. All three had been agricultural counties with relatively sparse populations until their local economies were transformed by the railroad and the coal industries. As Table 2 illustrates, in 1890, just as McDowell was poised to undergo industrial transformation, the then larger but still farming and timber dependent Logan had more residents. In 1895, Mingo barely possessed the 6,000 inhabitants necessary for the formation of a new county and that only after "considerable conniving." However, by 1900, with a population of 11,359, Mingo surpassed its parent county by almost doubling its population in just five years. By comparison, Logan County's population, which had been 11,101 in 1890, fell to 6,955 in 1900. Table 2 illustrates how the railroad and coal industries first precipitated the early growth of Mingo and the contemporaneous decline of Logan, and then, after finally coming to Logan in 1904, reignited development there.[53]

The degree to which Logan lagged behind because it was remote from the railroad and had no coal development is underscored by financial statistics from the 1900 Agricultural and Manufacturing Census. In 1900, Mingo County's "capital investment in manufacturing and mechanical industries amounted to $268,975, but in Logan investments only came to $9,285." The rapidity of Logan's development after being accessed by a railroad and "opened" for the development of its coal properties is the last significant illustration in Table 2. After the Chesapeake & Ohio rail line to Logan County became operational in 1904, Island Creek Coal Company began mining. Logan embarked on a transformation that by 1920 eclipsed Mingo.[54]

County	1890	1895	1900	1910	1920
Mingo*	N/A	6,000	11,359	19,431	26,364
Logan	11,101	5,101	6,955	14,476	41,006
McDowell	7,300	N/A	18,747	47,856	68,571

Table 2. Population of Mingo, Logan, and McDowell Counties, 1890–1920
Source: "Fourteenth Census of the United States, State Compendium: West Virginia." *Mingo County did not exist in 1890; it was divided from Logan County in 1895.

In addition to the influence of the railroads, the creation of the U. S. Steel Corporation in 1901 also significantly altered the course of national and regional economic development. U.S. Steel, the first billion-dollar corporation in the United States and the world, effectively dominated the export of southern West Virginia coal after it purchased almost 80 percent of the Pocahontas coalfield. As noted by the UMWA in 1921, U.S. Steel also eventually accumulated 50 to 60 thousand acres of coal lands in Mingo County.[55]

For the entire period between 1901 and 1920, Mingo County did not warrant the recognition afforded the U.S. Steel properties in the Pocahontas coalfield. Mingo's coalfield possessed a high quality coal, but it was of lesser quality and commanded less demand than the coal found in McDowell. McDowell's smokeless coal was considered of such high quality, demand for it offset the costs of operating in an older, less efficient coalfield. The production ranking of the two coalfields underscores the difference in the market value of their coal. Mingo County never placed higher than seventh among West Virginia's coal-producing counties, whereas neighboring McDowell County, older and less efficient, but the heart of the Pocahontas (smokeless) coalfield, consistently ranked either first or second.[56]

In 1904, when the Guyan coalfield replaced Thacker as the youngest coalfield in southern West Virginia, the peripheral status of Mingo's coalfield became permanent. Although both the Guyan and Thacker coalfields mined high volatile coal, development of the Guyan coalfield more closely resembled that of the smokeless coalfields. Just as in McDowell's Pocahontas coalfield, highly capitalized corporations dominated the development of the Guyan coalfield. The largest, Island Creek, a subsidiary of U.S. Coal & Oil, owned its coal lands in fee simple, or outright. As a result, "shoestring operators stayed away." Launched after the demise of unionization efforts in southern West Virginia, and free from the distractive presence of small-time would-be entrepreneurs, the operators of the Guyan coalfield ran their empire in Logan County smoothly for almost twenty years.[57]

The relationship between the operators and the railroad during the initial development phase of the Thacker and Guyan coalfields underscores the contrasting economic power of the two young coalfields. Andrew Roy had observed in 1899 that one of the few disadvantages faced by operators in the Thacker coalfield was their weak bargaining position with the Norfolk & Western. As Roy noted, because the N&W had no competitors for trade in the Thacker coalfield, it required the coal companies to build their own switches. Just five years later when the Chesapeake & Ohio lagged in constructing connections to the new Island Creek mines, the company threatened to renew its own plans to build a connecting line to the N&W tracks at Canterbury, in Mingo County. Eager not to lose out to the N&W, the C&O completed the work and granted favorable rates to Island Creek. Five years after starting production, Logan's Guyan coalfield outranked Mingo's Thacker coalfield.[58]

In an anecdotal history of mining in southern West Virginia, pioneer coal operator W. R. Thurmond discussed important differences between the six coalfields of southern West Virginia. Most important, Thurmond also asserted that historians all too often conceptualized all six coalfields as a single unit. According to Thurmond, "the basic division . . . is an east-west one." In eastern southeastern West Virginia were the three smokeless coalfields: the New River, Pocahontas, and Winding Gulf; in the west were the high volatile coalfields: Kanawha, Logan, and Williamson. Thurmond believed the differences among the six coalfields extended further than the type of coal they produced.[59]

According to Thurmond, the three eastern coalfields could be thought of as a "unit" because they "had much in common: capital, markets, la-

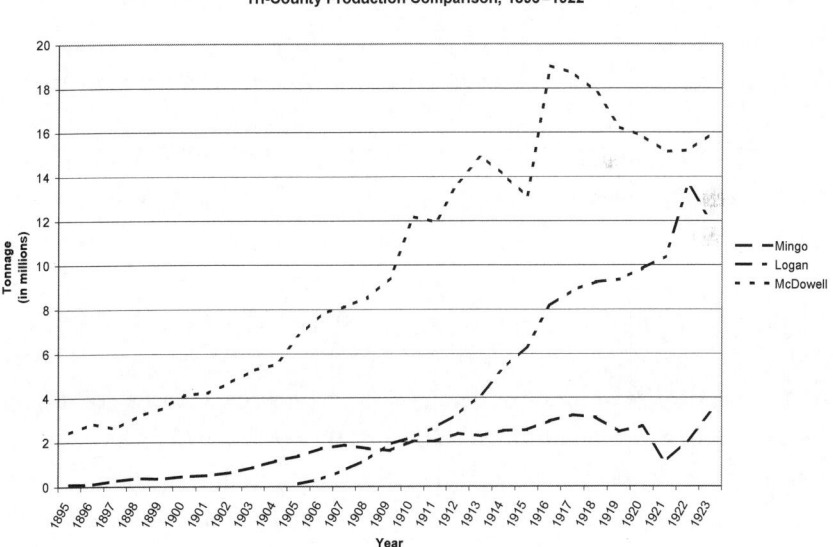

Tri-County Production Comparison, 1895–1922

Coal Production in Mingo, Logan, and McDowell Counties, 1895–1922. *Mingo's coal production paled in comparison with its larger, better capitalized neighbors.* *["Annual Reports" of the West Virginia Department of Mines.]*

bor relations, traditions and personnel." By contrast, "the three western coalfields developed more or less independently of each other." Thurmond's most telling observation, which stands in marked contrast to most scholarly treatments of southern West Virginia coal history, is his claim that "the western fields were not developed as a mere extension of the older eastern fields." Further, Thurmond states that "the operators in the Smokeless area played on a different stage . . . reading from a different script."[60]

Although Thurmond's analysis of the development of the six southern West Virginia coalfields emerges from armchair reminiscence, most of his commentary can be substantiated. Moreover, the differences he stressed did contribute significantly to the eruption of labor violence in Mingo County. First, although Thurmond's assertion about the cohesion in the eastern coalfields can be carried too far, it is worth noting that even though they competed against one another, the smokeless operators all benefitted from the market value of the smokeless label. The western high volatile operators enjoyed no parallel perk.[61]

Second, as Thurmond states, the operators of the smokeless coalfields were a tightly knit group. Several were active members in more than one of the smokeless coalfields' operators' associations. While the Kanawha and Logan coalfields were similar to the smokeless coalfields in that there were operators with connections to both coalfields, the Williamson-Thacker coalfield was unique. Many of the investors, operators, and employees of mining companies of the smokeless and high volatile coalfields also invested in the Williamson-Thacker coalfield. However, as mentioned before, investment did not involve active, on-site participation in the running of the mines.[62]

The first decade of the twentieth century witnessed a pattern of expansion and contraction that ultimately limited the coal industry's horizons in Mingo County. The opening of a new N&W branch line in 1902 focused attention on the concentration of coal development in just three districts; 21 of Mingo's 25 mines were still located in Lee, Magnolia, and Stafford districts in 1904. The other four mines, located in Harvey District in the northwestern corner of the county, were operated at Dingess by the Pearl Coal Mining Company, one of the oldest companies in the coalfield. The relocation of the railroad in 1902 and the economic downturn of 1904

The Six Coalfields of
Southern West Virginia

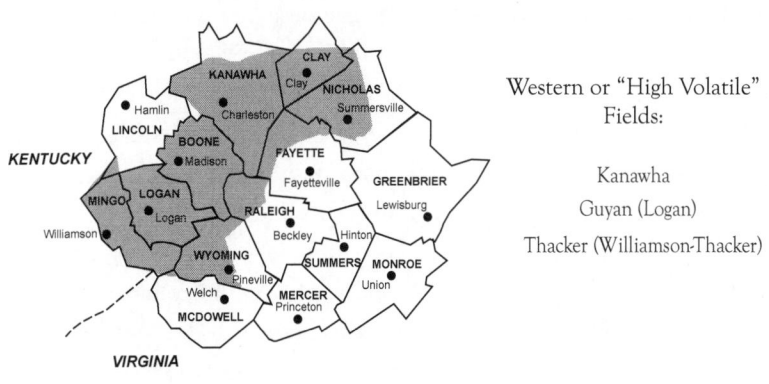

Western or "High Volatile"
Fields:

Kanawha

Guyan (Logan)

Thacker (Williamson-Thacker)

Eastern or "Smokeless"
Fields:

Winding Gulf

New River

Flat-Top Pocahontas (Pocahontas)

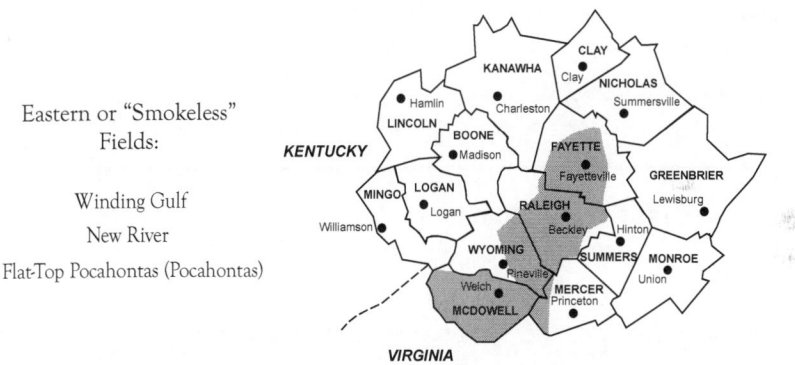

The Coalfields of southern West Virginia. *The western "High Volatile Fields" and the eastern "Smokeless Fields." [Created by E-S Bonita Bonar]*

spelled the end of Dingess as a mining center, and eventually even the train tracks were removed. From 1902 through the extended recovery from the 1907 "Bankers'" or Copper Panic, it became harder for small, entrepreneurial mine operations to survive, much less grow. The collapse of the U.S. copper market on October 21, 1907, had sparked a general stock market crash and "a sharp financial panic." As described by historian Frederick Barkey, in West Virginia the "so-called 'Bankers' Panic' was anything but a short economic set back," and small businesses went under daily. Just the latest in a long line of United States financial panics, which recurred "with worrisome regularity every ten years" in the nineteenth and early twentieth centuries, the 1907 panic precipitated a contraction of the coal industry.[63]

Generally, during an upswing in the business cycle, the price for coal followed demand upward and the number of coal companies in a coalfield increased. Many of these companies were poorly capitalized, high risk ventures undertaken by investors hoping to make quick money and get out. Other small-time capitalists were investors who were unable to raise sufficient capital to open a coal mine in a tight economy but benefitted from the easier banking standards of an inflationary cycle. In either case, when the demand for coal decreased and/or price deflation set in, these companies shut down. The larger, better capitalized companies would then move in and buy out these flagging operations.[64]

The contraction and consolidation of the coal industry in Mingo County that followed the 1907 panic mirrored the state and national reaction to the recession. Twenty-seven companies operated 42 mines in Mingo County in 1907. Just two years later, there were only 19 companies running 32 mines. The depression of 1908 also contributed to the general economic constriction which had begun in 1907.[65]

In 1909 two developments in the Williamson-Thacker coalfield continued to mirror national and/or state trends. More important, they marked the beginning of the end of one cycle in the long transformation of the coal industry from a nineteenth-century industry to a twentieth-century industry. As mentioned earlier, in the coal industry, as in the marketplace at large, economic downturns forced the closure of small and marginally capitalized companies. During the bust or in its aftermath, larger and/or more financially secure operations often exploited a deflationary lull in the demand for coal lands.[66]

The emergence of large multi-state coal corporations and the decline of entrepreneurial owner-operator companies began before the turn of the century and was marked in West Virginia by the underwriting of development by corporations from Pennsylvania, Maryland, and New York.

The consolidation of the rail industry begun by J. P. Morgan and others in 1898, and the creation of U.S. Steel in 1901, accelerated this transformation. Many contemporary observers, and especially the "corporate" coal men, argued that their domination of the industry meant progress because inflationary and cutthroat competition would be diminished, which in turn would enable their companies to focus on efficient and safer coal production.[67]

Mechanization of the extraction process also diminished the need for a skilled and demanding work force. The traditional work process was "restructured—piecework fell away, hour work increased, task specialization, [and] supervision increased." According to coal historian Keith Dix, 1909 marked the turning point from traditional to modern mining in West Virginia. Dix based this assertion on an examination of work-related statistics, the most important being the ratio of the "skilled craftsman" pick miners to machine miners. The advent of machines reduced the importance of the individual man in the mining process. Miners devolved from quasi-autonomous subcontractors into just another type of interchangeable part.[68]

As illustrated by Table 3, employment statistics from Mingo County mirror the state trend noted by Dix. As the number of mines in the Thacker or Williamson-Thacker coalfield increased, so too did the number of pick miners. However, because machine mining had accompanied the coalfield's opening, there was a corollary increase in the number of machine miners. In Mingo County, machine miners surpassed pick miners three years before the state average. Mechanization also made possible the division of tasks in and around the mines, which meant that the number of men employed by the mines, but not as pick or machine miners, also increased. The table also illustrates this trend. The change was gradual and was influenced by many factors. For example, the continued decline in the number of pick miners after 1906 was accompanied in the 1908 depression year by an overall decline in the total number of men employed.

In 1910, Mingo's coal production passed the two-million-ton mark. The Thacker Coal & Coke Company alone worked ten mines. The consolidation of larger mines continued as the Magnolia Coal & Coke Company absorbed the old Mate Creek operation. However, neither increasing coal production nor advancing technological sophistication could protect Mingo's miners from misfortune or prevent the coalfield's marginalization, as illustrated by the Lick Fork mine disaster.[69]

Year	Total # of Mine Employees	Pick Miners	Machine Miners	% Pick Miners
1898	673	329	105	49
1899	617	352	52	57
1900	751	371	148	49
1901	1,414	655	349	46
1902	1,489	710	395	48
1903	1,172	615	229	52
1904	1,781	666	357	54
1905	2,548	786	708	31
1906	2,624	574	755	22
1907	2,624	615	784	23
1908	1,677	412	717	25
1909	2,053	245	785	12
1910	2,358	241	745	10

Table 3. A Comparison in the Number of Pick and Machine Miners in Mingo County, 1898 to 1910* *Source: "Annual Reports of the West Virginia Department of Mines, 1898–1910." *Data prior to 1898 were not available.*

Tragedy struck at the old Lick Fork mine shortly after 7:00 a.m. on December 31, 1910. The ten men who had worked the night shift were in the first of four cars on the upward-moving man trip when the last three cars broke loose. The loss of the back three cars upset the counterbalance between the man trip and the cars traveling down slope. The car with the men came up the 1200-foot slope so fast that when it hit the headhouse, seven of the men were killed instantly, and the others died later from their injuries.[70]

The rocketing shot of the Lick Fork man trip symbolized in a sense the growth of Mingo County and the Williamson-Thacker coalfield. Despite enjoying increased coal production, population, and economic expansion, Mingo County failed to keep pace with its neighbors. The average value of land per acre in Mingo had risen 47 percent between 1900 and 1910, from $7.57 an acre to $16.28. Comparison of the same statistics for Logan and McDowell Counties reveals that Mingo's land value, which in 1900 had been comparable to McDowell's and ahead of Logan's, lagged in 1910. Table 4 illustrates this growing disparity in land values.[71]

Possessing the premier coal in the state, McDowell remained West Virginia's top producer, and as the virtually captive source of coal for U.S. Steel, had secure sales. Opened only in 1904, Logan County mining owed its startling growth to an aggressive development by highly capitalized operations.[72]

Year	Mingo	Logan	McDowell
1900	$7.57	$5.76	$8.04
1910	$16.28	$32.03	$33.42

Table 4. Comparison of Land Values in Mingo, Logan, and McDowell Counties, 1900–1910 (in $ per acre) *Source: "Abstract of the Thirteenth United States Census, with Supplement for West Virginia."*

1911 proved to be an important turning point for the coal industry in Mingo County. Over the course of the year, two cases marked a separation between the coal interests of Mingo County and the interests of the counties to the east. The first event centered on a conflict between the Williamson-Thacker coalfield and the Pocahontas coalfield over an unfair market advantage. The second case focused on the break between one of Mingo's largest coal operations and the Norfolk & Western's official coal sales agency. The backdrop for these events was the tense condition of the American economy. 1911 was a low point in the national economic cycle, which marked the turn toward recovery.[73]

Basic similarities between the Thacker coalfield and surrounding coalfields in their early development allowed historians to overlook differences that took years to take effect. For example, the coal mined during the Thacker coalfield's first decade was shipped east, like the smokeless coalfields' coal, for use by steamships. However, the UMWA strikes and unionization drives of the late 1890s and early 1900s in Pennsylvania and the Midwest allowed the coalfields of southwestern West Virginia to capture an increasing share of the Great Lakes coal trade. The split in the export direction of southern West Virginia coals eventually contributed to greater competition between the southern West Virginia coalfields. Haulage rates for these coalfields differed even when the coal traveled in the same direction on the same railroad. For example, high volatile coal from the Williamson-Thacker coalfield and smokeless coal from the Pocahontas coalfield were both shipped to the Great Lakes region on the Norfolk & Western Railway. Despite being neighboring coalfields, and despite the Pocahontas coalfield being farther away from the coal's destination, the Williamson-Thacker coalfield was designated an "inner crescent" coalfield and paid a lower haulage rate than the Pocahontas coalfield, which was categorized as an "outer crescent field."[74]

Market competition was always a source of tension among the coalfields of southern West Virginia because they competed for market shares not only against each other but also with coalfields from the Mid-Atlantic through the Midwest. Coal from southern West Virginia was transported to the shipyards of Tidewater Virginia and to the Midwest. The western coal route was referred to as the Lake Cargo traffic because the coal shipped on the railroads that followed the Ohio River system traveled all the way to the steel-producing centers along the Great Lakes.[75]

Although the N&W coalfields shipped coal to the Great Lakes as early as 1903, that area did not become the primary market for Mingo County coal until 1921. However, from 1911 until 1921, as much coal from Mingo traveled to the northwest as was shipped east. Superficially, this shift appeared to liberate Mingo's coal mines from a competition they could not win against the smokeless coalfields to the east. In reality, the shift squeezed Mingo even more, and heightened corporate tensions between the Williamson-Thacker and Pocahontas coalfields.[76]

The source of these adversarial attitudes was a geographic-economic division among the coalfields that contributed to the Lake Cargo traffic. The Williamson-Thacker coalfield in Mingo County was the southernmost West Virginia inner crescent coalfield, while its closest neighbor, the Pocahontas coalfield in McDowell, marked the beginning of the outer crescent. As an inner crescent coalfield, Mingo paid a lower shipment rate than the McDowell coalfield. To make matters worse, the operators of West Virginia's N&W coalfields who shipped their coal over the Pennsylvania Railroad system paid lower rates than the Pennsylvania operators. Tension over this rate differential resulted in federal hearings in Washington in March 1911, which pitted the inner and outer crescent coalfields against each other as well as against the operators from Pennsylvania.[77]

Despite the agitation precipitated by the rate differential controversy, the Mingo County coal operators committed themselves to pursuing the western and northwestern markets. On March 24, 1911, the *Mingo Republican* reported that the Sycamore Coal Company celebrated its first shipment of six coal cars from its new plant to Cincinnati by arraying the first car with ceremonial bunting. Also in 1911, Borderland Coal Corporation ended its coal sales contract with the Leckie Coal Company, a coal sales agency in Cleveland, Ohio, and opened its own agency based in Cincinnati. The Borderland coal sales agency eventually also served as agent for two other mines in Mingo County.[78]

The decision to concentrate on the western and northwestern markets was probably not Borderland's only consideration in establishing a coal sales agency. The Cooper family coal mines in the Pocahontas coalfield also shipped to Ohio. In establishing their own agency, the Coopers claimed to be motivated by "the advantages to both consumer and producer resulting from direct dealings" between the two. The final factor that might have inspired both Borderland and the Coopers to establish coal sales agencies was the 1911 antitrust suit filed by the Taft Administration against U.S. Steel. The U.S. Steel suit "miscarried," but the assault on the giant which virtually controlled the Pocahontas coalfield created a window of opportunity for companies seeking to establish a degree of autonomy from the corporate leviathan they feared would focus anti-monopoly sentiment against them.[79]

The relative weakness of the coal industry in the Thacker coalfield also affected political and social relations in Mingo County. The industrial elite influenced politics in the county, but not to the degree of their contemporaries in other coalfields. And, although industrialization transformed the social landscape of Mingo County, it did not utterly sweep aside patterns of behavior that predated the advent of commercial coal mining. The persistence of traditional political dueling and marginal modernization set Mingo County further apart from its neighbors.

During Mingo's first two decades, industrial development precipitated profound political and economic changes, which, in turn, affected community growth and social relations. Conversely, both the nature and limits of these changes meant that the new "industrial order" failed to eradicate traditional beliefs and patterns of behavior. Possessing at best a tenuous grip on political power and distracted by chronic economic instability, Mingo's coal elite was forced to coexist with, and adapt to, a frequently defiant local culture.

Obscured by later stories of how miners and their families "owed their souls to the company store," early on in Mingo's history natives flocked to emerging towns and coal communities as a promising alternative to an increasingly bleak future. Agricultural statistics from the 1910 census expose one part of the new reality they faced. Although the number of farms had increased by 50 percent in just 20 years, the amount of farmable land had decreased by 11 percent. The farms were smaller, less productive, and more likely to be worked by tenant farmers or share-croppers. Compounding the farmers' plight was a crisis in soil erosion caused by the expansion and intensification of agriculture precipitated by forest clearance. By 1897, when the Army Corps of Engineers reported the soil erosion problem on the Tug, the era of entrepreneurial logging was ending. There were still thousands of acres of trees standing upstream on the Tug Fork's principal tributaries, but they could only be exported by those with the capital to transport them by rail. Thus, the residents of the Tug Valley were well on the path to economic dependence on the coal industry within a decade of its arrival.[80]

With little alternative in the modern cash-driven economy, the native population turned increasingly to work in the mines. One Mingo amateur historian noted that for the mountain people living in cabins with holes in the chinks "big enough to sling a dog through," work in the mines that were "[popping] up overnight, like mushrooms after a rain" did not seem that unwelcome. However, the lack of opportunity for other employment did not make Mingo natives an entirely biddable workforce.[81]

The oral testimonies from Mingo County reveal that even though the residents' dependency on the coal industry increased, they devised a variety of economic strategies to retain as much independence as possible. Kinship rooted most forms of resistance. When out of work, the landless survived because they could set up temporary residence with land-owning relatives. Although 94 to 98 percent of southern West Virginia miners resided in company housing, this figure obscures the significance of independence to those who did not. Testimonials from the Matewan Oral History project 1989–1990 abound with variations of the declaration, "we always lived in our own house, on our own land." When labor strife erupted in Mingo in 1920, the strike's most visible local leaders were not culturally rootless transients alienated by the company system, but natives who had adamantly refused to live in company housing.[82]

Close ties between Mingo's mining and farming populations were also seen as autonomy-preserving activities. Farmers peddled all manner of goods and services through the camps, including their surplus vegetables, fruits, and meats. These transactions helped keep miners from complete dependence on the company store, and also diverted some money into the local economy. Many a farming family's income was supplemented by the mining wages of unmarried sons, or by men who combined farming with seasonal work in the mines. Unfortunately, in the long-term, the combination of mining and farming contributed to the impoverishment of the local economy. By accepting lower wages for what they deemed supplemental labor, miner-farmers helped keep wages depressed.[83]

Coal camp life and the relations between miners and operators in the "company system" have dominated the social history of southern West Virginia. In Mingo County, however, mining operations and independent communities often occupied the same location. The story of Dingess, an amalgamation of company and independent town, encapsulates both the dependence of the local economy on the coal and railroad industries and the persistent complexity of political and social networks in Mingo County. In 1891, the year Wallace J. Williamson and the Williamson Mining and Manufacturing Company founded the town of Williamson, Dingess was already a village of 100. Like Williamson, Dingess grew rapidly after becoming the population center for three of the county's first mines. Within two decades, however, Dingess's population had fallen to 75, and the town had become a haven for crime and social "degeneracy."[84]

At its peak, in 1903, the town of Dingess had 750 inhabitants, which made it second in population only to the county seat Williamson. In population it surpassed both Thacker and Matewan, respectively the center of the largest coal operation in the county and the political rival of Williamson. In 1904, in recognition of Dingess's growth, J. R. Booth, the superintendent of the Pearl Coal Mining Company, was appointed to the Executive Committee of the county Republican organization. Booth's ascendance seemed to indicate that political power and economic influence had broken free of the Williamson-Matewan, Lee-Magnolia orbit and expanded to incorporate the county's remote areas. However, 1904 proved to be a high water mark in the long decline of Dingess's history as a community.[85]

In contrast to Thacker, Red Jacket, and Glenalum, which sprang up and continued to grow, Dingess, which had 200 miners at work in 1893, began to deteriorate rapidly after only a decade. There were two primary and inter-related reasons for this situation. First, the coal seams were thinner and of inferior quality at Dingess. Second, once a new section of the county was accessed by the railroad in 1904, maintaining the high-maintenance tracks to Dingess was no longer cost effective. Within a decade the mines there were abandoned and the population dwindled to its pre-development level.[86]

Actually, Dingess declined even more rapidly than the surveyors realized. The town's population had fallen to one-tenth of its 1902–1903 level in less than five years. With the mines closed, the mobile mining population dispersed, and the rail line fallen into disrepair, the native inhabitants were left behind to fend for themselves. However, Dingess retained political influence far out of proportion to its size. As Mingo's own rotten borough, Dingess offered the machine politicians of the county a secure bloc of votes in return

Dingess, West Virginia. *This photograph captures the town's decline after the mines closed. [WVRHC]*

for a steady flow of funds and patronage plums. By World War I, in recognition of its now dubious reputation, Dingess was identified as the center of Mingo's "wild and rough" region. Local citizens were encouraged to think of Dingess as the ultimate example of Mingo's pre-industrial backwardness.[87]

Industrialization wrought significant changes in Mingo County. As documented in Tables 5 and 6, communities prospered or deteriorated depending on access to the railroad and the opening of mines. Although farming and logging continued after their arrival, fewer families found these activities economically sustaining. Mingo's inhabitants were drawn inexorably into the nexus of commercial mining. The destabilization of traditional living and occupational habits applied stress to familial and social ties. As a result, both opportunity for financial gain and the struggle for survival often overwhelmed customary behavior and not infrequently resulted in violence.[88]

Towns in 1891–92	Population	Towns in 1895–96	Population
Burch	150	no listing	
Dingess	100	Dingess	200
Eugene	100	no listing	
Nolan	25	Nolan	75
Spaulding	25	no listing	

Table 5. Town Growth and Decline in Mingo County, 1891–1896
Source: "West Virginia State Gazetteer and Business Directory, 1891–1892 and 1895–1896."

As an example of what this kind of economic stress can do, consider the tale that haunts Matewan of a family torn asunder by a son's greed. Through marriage to Phoebe, the sister of Devil Anse Hatfield, Anderson Ferrell acquired ownership of the farm that became Matewan. After Phoebe's death, Ferrell married Mary Chambers and started a second family. Seeking to provide financial stability for all of his children by capitalizing on the railroad's proximity, Ferrell divided his land. To his children by Phoebe Hatfield, Ferrell deeded small farms, and he sold the remaining land in lots, thus founding the town of Matewan. Among the entrepreneurs who purchased property from Ferrell was his nephew by marriage E. B. Chambers. Years after the Hatfield-Ferrell farm had grown into a thriving village, one of Anderson Ferrell's sons from his first marriage challenged his father's dispensation of the land.[89]

New Towns in 1895	Population	New Towns in 1895	Population
Breeding/Breeden	150	Sheppard	75
Edgarton	25	Thacker	400
Fairfax	100	Wharncliffe	12
Hinch	150	Williamson	500
Matewan	200		

Table 6. Mingo County Towns that Had Not Existed in 1891 and their Populations in 1895–1896 Source: "West Virginia State Gazetteer and Business Directory, 1891–1892 and 1895–1896."

Aided by Henry Clay Ragland, the attorney who had originally served his illiterate father, Floyd Ferrell brought suit to gain a larger portion of the family property, which now encompassed the town of Matewan. When the local circuit court judge granted his son's claim, Anderson Ferrell appealed to the West Virginia State Supreme Court of Appeals. The court chastised the younger Ferrell by asking, "What right had this son to the property?" when it would lead to the bankruptcy of his father. The court also questioned Floyd Ferrell's motives in pursuing the suit, which at one point he had withdrawn in order to "let the second wife and little children of his aged father have something for home and bread." The court extended its reprimand to include lawyer Ragland for encouraging Floyd Ferrell to take action against his father. The acrimony generated by *Ferrell v. Ferrell* survived for decades and may have contributed to a rivalry between the Hatfields and the Chambers, who struggled against each other for decades for control of Matewan.[90]

Adaptation to the economic transformation of the region not only affected relations within native families; it also heightened tensions between native-born whites and other ethnic groups. Although the percentage of native whites in the county population only fell from 96 to 86 percent between 1900 and 1910, the influx of African Americans and foreign-born whites coincided with difficult economic times. In the aftermath of the Bankers' Panic of 1907, mines in the Thacker coalfield were closed or absorbed by larger companies, and miners were forced to compete for fewer jobs. Because of the high percentage of local whites in the Thacker coalfield, they viewed all migrant miners, but especially the foreigners and African Americans, as interlopers who were stealing jobs that rightfully belonged to the natives who were there first. Although episodes from the 1920–1922 strike period indicate an evolution toward cross-racial and eth-nic working-class solidarity, an episode from 1908 exemplifies the impact of chronic economic instability on heterogeneous mining communities.[91]

At the Pike Colliery mines across the river from Matewan, in Kentucky, a group of men attacked the home of two Hungarian families. In the course of committing what the *Williamson Enterprise* called "a most revolt-ing and brutal crime," the assailants injured the children so badly that their recovery was not expected. Five men were convicted for shooting and raping members of the two families and received sentences ranging from five to ten years. Neither the reason for the crime nor the exact nature of the attacks on the children were specified in the press account of the story. However, one detail provides insight into Mingo County's polarized reaction to a later event. One of the convicted assailants, Van Clay, reap-pears in Mingo County public records in 1920, as one of the leaders of the miners' strike. Probably unaware of the earlier incident, most historians have dismissed contemporary condemnation of Mingo's striking miners as "trash," or "criminals" as elitist snobbery. The extralegal and/or crimi-nal excesses of the miners have been excused as a reaction to the environ-ment created by the coal companies and part of the price of fighting for justice. One historian has even described the miners' violence as "politics by other means."[92]

The revelation about Van Clay's previous criminal record requires a mild adjustment of the historical analysis of both Mingo's miners and local perceptions of the 1920–1922 strike. When his prison term expired, Clay returned home to a small community with a long memory. Whether or not he was "reformed" would not have mattered to a significant percentage of his neighbors. As in many Appalachian communities, toleration or censure of criminal behavior in Mingo County depended on a collective assessment of the motive. While most residents of Mingo County refused to judge others' bootlegging or commissions of "justifiable" homicides, they considered attacks on children unforgivable. For many residents, Van Clay's prominence in the miners' struggle in Mingo would have undermined its legitimacy. Although Clay's 1908 criminal record should not be allowed to cast a shadow over the entire miners' movement in Mingo, its impact must be considered in the evaluation of contemporary and subsequent local criticism of the strike.[93]

The peripheral status of Mingo's coalfield diminished the new coal elite's primacy in the county. One result of this can be found in the survival of local attitudes toward customary rights and violence. Three cases from the 1895–1911 period reveal not only how Mingo residents resolved conflict, but also how the community continued to define the law and justice according to its own standards. Moreover, these same cases expose the historical roots of the community's reaction to the Matewan Massacre.[94]

The first case, a land sale case that lasted seven years (1898–1905), illustrates why for many Mingo Countians there was no hypocrisy in the Massacre defendants pleading self-defense in their deadly attack on 12 Baldwin-Felts agents, only three of whom were armed. In January of 1898, a creditor of Sanford Hatfield forced the sale of his farm, which was purchased for its mineral deposits by R. C. "Clay" Allison. Allison, who purchased the farm for just over one dollar an acre, promptly sold a half interest to an outsider capitalist. Three years after the sale, Hatfield sued to regain full ownership of the land for which he still paid taxes. The local circuit court ruled in Hatfield's favor, but Allison and McGee challenged the ruling. Four years later, or seven years after the original sale, West Virginia's State Supreme Court of Appeals overturned the lower court's decision and upheld Allison and McGee's ownership of the land. Testimony and case filings from the Supreme Court report recount the methods used by Hatfield and Allison in their attempts to manipulate the legal system.[95]

To Sanford Hatfield, the case boiled down to Clay Allison's failure to keep his word; according to Hatfield, Allison purchased the land in trust for him. Angered by Allison's repeated refusals to redeem the farm, Hatfield decided to sue. Hatfield later claimed that he decided to bring suit because he noted at the time of the sale the land was worth \$2,500 and "if it had not been understood that the land was being bought for him, it would have gone for three times what it did." Hatfield based his case on the oral testimony of his mother, father, sister, and Uncle Jim Hatfield. Ironically, Uncle Jim Hatfield contributed to the loss of the appeal, because in testimony, he refused to perjure himself on his nephew's behalf. Thus in contrast to the case made by Clay Allison and his attorney, the justices found Hatfield's argument "manufactured" and unbelievable.[96]

Clay Allison based his appeal to win ownership of the Hatfield farm on a counterclaim that Hatfield was seeking benefit, after the fact, from the farm's increase in value. A merchant for the community in which Hatfield lived, Allison presented numerous receipts and written documents illustrating Hatfield's debts and his own informal performance of banking functions. The judges were impressed both by Allison's possession of receipts, which documented his financial relationship with Hatfield, and the character of Allison's witnesses.[97]

The reasons the high court decided in Allison's favor despite noting inconsistencies in both men's version of events highlight the contrast between how the law was interpreted by residents of Mingo County and by the judges of the state's highest court. The local jury might have been swayed by Hatfield's desire to avoid dispossession and the former prominence of his family, but the final arbiter of the law in the state was not. Hatfield retained the surface rights to his farm, while Allison and McGee owned the mineral resources beneath the land. In return for not dispossessing Hatfield, his invalid wife, and several children, Allison and McGee did not have to pay taxes on the land, as that responsibility remained Hatfield's. The *Hatfield v. Allison* case underscores how the evolution of legal standards outside the Tug Valley affected social and economic relations within the Valley. While Sanford Hatfield had successfully manipulated the local court, his methods failed when a higher authority reviewed the case. The higher court's prejudice against Hatfield's reliance on familial witnesses, and the judges' propensity to believe the image of the "good citizen" businessmen summoned by Allison, reflects the divergence of attitudes toward family and social position within and outside the Valley. The jury's decision reflected customary values ignored by the high court. For example, Hatfield's fellow Mingo Countians probably turned a blind eye to perjurious testimony because they believed that it was unjust for Hatfield to pay taxes on land that Allison enjoyed the benefits of developing. By contrast, the Supreme Court judges, predominantly proponents of the new order, were swayed by Allison's mustering of evidence of contractual rights.[98]

If the judges of the West Virginia State Supreme Court of Appeals exhibited little patience with the attempt by Mingo Countians to protect Sanford Hatfield's farm, they displayed open disgust and disdain for the community's failure to contain the Hatfields' propensity for violence in the appeal of the *State of West Virginia v. Elias Hatfield* case. On July 3, 1899, Elias Hatfield, the fifth son of Devil Anse Hatfield, shot and killed Humphrey E. "Doc" Ellis at the train yard at Gray, West Virginia. Most sources agree that the reason for the deadly confrontation was Ellis's kidnapping of Elias's brother Johnse, whom Ellis had turned over to Kentucky authorities. Despite acknowledging that he sought out Ellis, Elias Hatfield pleaded self-defense and justifiable homicide.[99]

Unfortunately for Elias Hatfield, Doc Ellis was a popular man. Public outcry led to Hatfield's arrest and trial; the demand for justice was so strident that West Virginia governor George W. Atkinson personally took him into custody for safekeeping. However, the sentence Elias Hatfield received stirred even more public enmity. Convicted of murder in the second degree, Elias Hatfield was sentenced to just 12 years.[100]

A closer look at the Ellis-Hatfield murder case reveals yet another element in the pattern of the persistent traditional definition and application of moral justice in Mingo County. The two sources for this reevaluation are commentary from the majority opinion in Hatfield's appeal to the West Virginia Supreme Court of Appeals and "On the Circuit in Southern West Virginia," a 1901 article in the legal magazine *Greenbag*.

The majority opinion in *State v. Hatfield* gives voice to the conflict between local customs and the demands of modern law. After wryly commenting on the difficulty the lower court must have had in trying to impanel a jury *not* dominated by Hatfield kinsmen, Judge McWhorter embarked on a stern condemnation of the defendant *and* his community. In response to Hatfield's claim that he had killed Doc Ellis in self-defense, McWhorter mocked:

> What was [the] defendant doing with his Winchester rifle in his hands? . . . He had gone to take some letters to the post office, surely a peaceable mission. It was but a short distance from his place of business to the post office and not through a hostile section infested with wild beasts . . . [or] robbers and brigands.

McWhorter went on to express his disappointment with the community in which the confrontation occurred, saying, "It would seem that these reformers would have rid the country by this time of these dangerous characters, so that it would no longer be necessary to carry a Winchester constantly in self-defense when about the ordinary duties of life." With these words, Judge McWhorter proclaimed the end of West Virginia's legal complicity with frontier justice. It was unacceptable for a man to move about in polite society armed, much less for him to purposefully seek out a man intending to kill him, provoke the intended victim to attack, kill him, and then escape punishment by pleading self-defense.[101]

The notoriety of the Ellis-Hatfield case refused to fade. In 1901, Edwin S. Doolittle made the case the centerpiece of an article titled, "On the Circuit in Southern West Virginia." After describing the primitive traveling conditions endured by those who "rode the circuit," the author focused his observations on the antiquated but effective oratorical style of the lawyers. According to the author of the piece, former West Virginia Governor E. Willis Wilson's defense of Elias Hatfield stands as the paramount example of the traditional legal approach to defense. First, Wilson's two-hour long closing argument turned the victim into the villain. Second, the author noted, it was this use "of eloquent but irrelevant oral argument" and not the merits of the case that resulted in a victory for the undeserving.[102]

Despite the best efforts of legal and social reformers, for decades lawyers continued to use similar tactics in arguing before juries in southern West Virginia. In 1920, the lead attorney for the Matewan Massacre defendants, Harold W. Houston, declined the proffered legal assistance of the American Civil Liberties Union for fear of alienating the jury. When the case went to trial in 1921, Houston dramatically depicted the Massacre defendants as men protecting their homes and families from an invasion of lawless thugs. He entreated the jury to return the "defenders" of Mingo County to their little children who prayed nightly for their fathers to come home. According to news reports, Lawyer Houston's eloquence left women prostrate and sobbing and defendant Sid Hatfield openly weeping. In both cases, the defense attorneys' ploys worked. Elias Hatfield, although convicted of second degree murder, was sentenced to serve a mere 12 years; the Massacre defendants were acquitted.[103]

An incident that occurred between the murder of "Doc" Ellis and the Matewan Massacre highlights another element in the pattern of violent conflict resolution in Mingo County. The story of the fatal exchange between Thomas Chafin and the police chief and mayor of Matewan in 1911 shows that the story of a violent event involving public officials in Mingo County was rarely simple. These crises never had one, single irrefutable cause. Moreover, efforts to unravel their why and how always seemed to be complicated by the deaths of the principals and deliberate obfuscation by the deceased's partisans. In the wake of these incidents, members of the community divided over which version of the story they believed, uniting only when outsiders pushed to know "what really happened."[104]

On Wednesday, April 26, 1911, Tom Chafin, a miner employed at Red Jacket and a nephew of Devil Anse Hatfield, shot and killed W. R. Hoskins and Walter E. Musick, respectively the mayor and police chief of Matewan. According to the newspaper accounts of the incident, both Tom Chafin and Walter Musick's wife believed that Musick and Chafin's wife were having an affair. On the Sunday before the shooting, Mrs. Musick had gone to the Chafin home and compelled Mrs. Chafin to accompany her to Matewan where Mrs. Musick tried to have Mrs. Chafin jailed. At this point the stories presented in the two newspapers diverge.[105]

Published on April 28, the *Mingo Republican*'s (Republican) account purported to be based on the deathbed statement of Mayor Hoskins. Hoskins claimed that Musick, the son of former sheriff E. E. Musick, hoped to convince Chafin that no affair had occurred by taking a solemn oath that "he had not invaded his home." Chief Musick spoke pleasantly at first with Mr. Chafin, but the two ended up in the yard scuffling, whereupon Chafin pulled a gun. Seeing the gun, Hoskins hurriedly approached, but Chafin shot Musick twice in the chest whereupon Musick fell dying into Hoskins's arms. Hoskins, who was not armed, turned to flee but was also mortally wounded by Chafin. The *Republican* closed its story with a brief alternative to the mayor's dying declaration. In this unverified version, Hoskins and Musick went to Chafin's home to serve a warrant and upon arrival, as they stood on the threshold, Hoskins allegedly told Chief Musick, "Get Chafin or kill him." Fearing for his safety Chafin opened fire on the two men.[106]

The Democratic paper, the *Williamson Enterprise*, published a quite different and innuendo-ridden version of the events. The *Enterprise* claimed that Chafin had returned home and found Musick in the house alone with his wife. Chafin opened fire on Musick, whereupon Hoskins, who "just happened to be in the neighborhood," came to investigate and arrest the shooter and was also wounded. Chafin fled the scene but eventually surrendered to the authorities. The *Enterprise* reported that Chafin, accompanied by his brother, turned himself in because he feared being caught by the deceased's relatives who had gone looking for him.[107]

Convicted of voluntary manslaughter and ordered to serve two years in the state penitentiary, Chafin escaped from the local jail and again went on the run from the authorities. A posse chased Chafin to Mercer County where he was found dead at the foot of Pigeon Mountain. The lack of Mingo County court records denies analysts the opportunity to prove their hypotheses about the Chafin-Musick-Hoskins encounter. However, several influential details teased from the newspaper accounts of the story seem to indicate the existence of a pattern observable in other episodes involving public officials and violence in Mingo County.[108]

First, the differences in the presentation of the story by the *Mingo Republican* and *Williamson Enterprise* hint at a political issue lurking in the background. Musick and Hoskins belonged to the Republican leadership of the Magnolia district, while most of Mingo's Chafins, like their kinsman Devil Anse Hatfield, had never deserted the Democratic Party.

Little wonder, then, that the Democratic paper published the account that asserted that Chafin caught Musick in the act of adultery. In the unspoken judgment of the paper, the cuckolded Chafin rightfully punished the man who destroyed the sanctity of his home and abused his honor.

Second, there is the "woman as scapegoat" issue. Although to some the accusation of adultery justified Musick's death, whether the accusation was true remained in question. The equally partisan *Republican* acknowledged Mrs. Musick's suspicions, but by repeating in full the mayor's declaration that Chief Musick went to Chafin's home to plead his innocence, the paper implied that there might have been a misunderstanding. However, the recurrent use of women to explain or justify episodes of violence involving Mingo's officials suggests a purposeful innuendo of moral turpitude.

As Musick's case demonstrates, women, or relations with women, pre-date the most infamous case in the county's history, at least in the local iconography of the violent confrontations of Mingo's public officials. The murder of Williamson police chief John B. Maynard in 1918 by Chattaroy Constable Jesse Huffman, also allegedly resulted from a "bitter feeling" between the men over a woman. Maynard's recent defiant independence in the faction wars for control of Williamson's municipal politics, while acknowledged in the press account of his death, apparently was not considered as a possible cause for his demise. The specter of illicit passion reared again in June 1920, when Cabell Testerman's widow Jessie married Sid Hatfield less than two weeks after Testerman died in the Matewan Massacre. According to Baldwin-Felts documents and several trial testimonies, Hatfield either staged the confrontation known as the Massacre in order to kill Testerman, or at least capitalized on the ensuing confusion to eliminate his rival. To this day, in some circles, whether Mingo Countians believe or deny the story about the "love triangle" remains an important indicator of individual attitudes about the Massacre, strike, and unionization.[109]

The third recurrent issue first highlighted in accounts of the Musick-Chafin story is the lack of an unassailable reason for the fatal encounter. The mystery surrounding the role of Mayor Hoskins in the Musick-Chafin exchange intensifies upon examination of the two newspapers' explanation for his presence. The *Enterprise* derisively, but obliquely, observed that he was "in the neighborhood," while the *Republican* equally enigmatically alluded to a public reason for the confrontation when it noted that Musick and Hoskins, armed with a warrant, traveled together to Chafins' home. What breach of the law might Chafin have committed to bring Hoskins and Musick out of their municipal jurisdiction to his door?

Although Chafin's death silenced the public record, the reasons for his death wrap another layer around the story. Why would Chafin surrender, go to trial, receive a light sentence, only to go back on the lam? Neither the *Republican* nor the *Enterprise* offer any explanation, other than the one originally offered to justify his original surrender, which was that he feared retribution from the relatives of Musick or Hoskins, or both.

The unanswered questions in this case, when combined with the conflicting assertions regarding its cause, eerily portended future events in the county. Just five years later, Mayor Hoskins's brother John, who was constable of the Magnolia District, shot and killed a man in an encounter also fueled by manly honor. Like Chafin, Hoskins ran, and then returned for trial. Although he was acquitted, almost two years to the day of his victim's death, Hoskins's lifeless body was found in Red Jacket's Mitchell Branch mine.[110]

Mingo County's industrialization did not automatically transform community attitudes about the role of law in conflict resolution or the administration of justice. *Hatfield v. Allison* and the *State of West Virginia v. Hatfield* both document the persistence of what amounted to acceptable manipulation of the law. The persistence of a traditional political economy that loosely interpreted "truth" and "fact" allowed jury nullification to promote the administration of what was right over what was legal. In *Hatfield v. Allison*, Sanford Hatfield and members of his family apparently were not troubled by committing perjury in order to regain control of his land. Similarly, Elias Hatfield believed that he could escape punishment for the revenge killing of "Doc" Ellis by concocting a self-defense excuse.

Details from the accounts of the incident that resulted in the deaths of Matewan's chief of police and mayor in 1911 also expose a pattern of local customary reaction to episodes of violence. First, the deaths of the principals conveniently and permanently obscured an accurate estimation of what initiated the conflict. Second, because of the absence of an irrefutable explanation, partisans of the involved parties constructed a version of events that divided the community. Either Thomas Chafin murdered Chief Musick and Mayor Hoskins rather than face justice for an unknown offense, or he righteously punished an adulterer. There were parallels between the community's interpretations of the deaths of Musick, Hoskins, and Chafin, and the Matewan Massacre and the subsequent murder of Sid Hatfield.

Specifically, public opinion divided over whether Hatfield and his allies were championing a just cause or Hatfield unnecessarily escalated a confrontation in order to murder the husband of the woman he coveted. Enmeshed in all of these stories is the issue of a growing divergence in social attitudes within Mingo County. However, these differing attitudes toward the law and justice did not place members of the local community into easily divisible categories of traditional and modern, native or outsider, working class or elite.

Long before Sid Hatfield and Albert Felts squared off in what one agent later called their deadly "bluff game," the political, economic, and cultural factors that put them in the doorway of Chambers Hardware Store were in place. Chaotic, corrupt, and frequently violent, Mingo County acquired the appellation "Bleeding Mingo" a full decade before the Massacre occurred. Unwise political alliances combined with fitful economic development limited the influence exerted by the county's predominantly "middle-manager" coal elite. Despite the sweeping social and cultural progress that industrial development was alleged to bring along with it, in Mingo, traditional views of what was "just" withstood more abstract and modern interpretations of the law. Instead of resolving these impediments to the relative quiescence "enjoyed" in the coal-dominated counties, the Progressive Era and World War I only exacerbated their influence and raised the stakes of the conflict.

The pressure in Mingo was building.[III]

NOTES

I   Although Mingo County was not created until 1895, the first mine in what would become Mingo was incorporated in 1891.

2   *Logan Banner,* October 20, 1898, as quoted in Edwin A. Cubby, "Transformation of the Tug and Guyandot Valleys, Economic Development and Social Change in West Virginia, 1888–1921" (Ph.D. diss., Syracuse University, 1962), 188.

3   By 1920 Mingo County's coalfield was most commonly referred to as the Williamson-Thacker coalfield. Initially it had been called the Thacker coalfield, after the coal and first major operation. Dr. Henry M. Payne, "The Future of Williamson and the Tug River Coal Field," *The Illustrated Monthly West Virginian* 7 (August 1908): 45–49; John Gaventa, *Power and Powerless: Quiescence and Rebellion in an Appalachian Valley* (Urbana: University of Illinois Press, 1980), 144.

4   Otis K. Rice and Stephen W. Brown, *West Virginia: A History.* 2d ed. (Lexington, KY: University Press of Kentucky, 1993), 208; John Alexander Williams, "The New Dominion and the Old: Antebellum and Statehood Politics as the Background of West Virginia's 'Bourbon Democracy'" *West Virginia History* 33 (July 1972): 391.

5   Sidney B. Lawson, *Fifty Years a Country Doctor: Autobiography and Reminiscences of Sidney B. Lawson, M.D.* (Logan, WV: n.p., 1941), 47–48.

6   Nancy Sue Smith, *An Early History of Mingo County, West Virginia* (Williamson, WV: Williamson Printing Co., 1960), 8–9; Altina L. Waller, *Feud: Hatfields, McCoys and Social Change in Appalachia, 1860–1900* (Chapel Hill: University of North Carolina Press, 1988), 154; "Incorporation of the Williamson Mining and Manufacturing Company," in *Acts of the Legislature of West Virginia,* Twentieth Regular Session (Charleston, WV: Moses W. Donnally, 1891), 88; *Logan Banner,* 16 July 1891; Phil M. Conley, "The Founder of the City of Williamson," *West Virginia Review* 2 (February 1925): 162; For more on the county seat war phenomenon see Ronald L. Lewis, *Transforming the Appalachian Countryside: Railroads, Deforestation, and Social Change in West Virginia, 1880–1920* (Chapel Hill: University of North Carolina Press, 1998): 215–234.

7   "Act Establishing Mingo County," *Acts of the West Virginia Legislature,* 1895, 213–214. The three commissioners were: J. K. Anderson, J. L. Deskins, and Alex Stafford. Source for naming of first sheriff: Smith, 8–9.

8   Undated "Bleeding Mingo" newspaper articles circa 1908 election, Roy H. Keadle Papers, West Virginia and Regional History Collection, West Virginia University. The bulk of this small collection consists of newspaper clippings, correspondence, and political and social ephemera, which were pasted into a ledger book.

9 *Mingo Republican,* 30 July 1914; *Logan Banner,* February 27, 1895, quoted in Cubby, "Transformation," 172; 1891–1892 *Gazetteer and Business Directory;* 1895–1896 *Gazetteer and Business Directory.*

10 1895–1896 *Gazetteer and Business Directory.* Williamson's population was 500; Thacker's was 400; Mingo County Court Commissioners' Record Book no.1, page 173, quoted in Cubby, "Transformation," 72, note 72; Initially Williamson was part of Lee District but was eventually redistricted independently. From 1895 through 1920, in population and coal employment figures, Williamson/Lee District and Matewan/Magnolia district were roughly equivalent in size.

11 Cubby, "Transformation," 171.

12 The most infamous case would be the Matewan Massacre, which occurred one week before the 1920 primary election. When in Matewan, it was abnormally quiet and voter turn out was small, the *Williamson Daily News* suggested that perhaps the two events were related: *Williamson Daily News,* May 27, 1920; Edgar P. Rucker to A. B. White (letter #7981), A. B. White Papers, WVRHC.

13 Source for exchange between Bias and Scott: Undated, unknown newspaper article in Keadle Papers, WVRHC; Source for profile of Cook: *Logan Banner,* September 13, 1894 article reprinted in *Charleston Daily Mail,* September 21, 1894.

14 The names of the Republican factions were given here because they can be documented in the sources cited for events discussed in this chapter. Ascertaining the names of the Democrats' factions was not possible until after 1911 when records on Democratic factional politics could be documented.

15 Ms. Margaret Ann Hatfield to author, letter no.17, author's possession. For nearly a year, from 1997 until Ms. Hatfield's death in 1998, she corresponded with the author who has kept and numbered Ms. Hatfield's letters. A wealth of Hatfield family and Matewan history, this collection of letters will one day be made available to the public. Hereafter referred to as "Hatfield correspondence" and the number of the letter; "G. T. Blankenship, political advertisement," *Williamson Daily News,* April 8, 1916; Connection between the Whites: Herman L. Fairchild, "Memoir (Memorial) of Israel C. White," (n.p., n.d.), 127.

16 The seven cases were: (1) *Williamson, et al v. County Court;* (2) *Hurst, et al v. Same;* (3) *Stafford, et al. v. Same.* (These cases were heard together). *Reports of the West Virginia Supreme Court of Appeals,* 56 (June 14, 1904): 38–43. (4) *Stafford v. Board of Canvassers, Reports of the West Virginia Supreme Court of Appeals,* 56

(June 14, 1904): 670–675; (5) *Stafford v. Sheppard, Reports of the West Virginia Supreme Court of Appeals,* 57 (January 24, 1905–April 25, 1905): 81–90; (6) *Stafford v. County Court, Reports of the West Virginia Supreme Court of Appeals,* 58 (April 25, 1905–February 6, 1906): 88–94; and (7) *Williamson v. Musick, Reports of the West Virginia Supreme Court of Appeals,* 60 (April 24, 1906–November 27, 1906): 58–75; Citation for the duration of the controversy was derived by mathematical deduction, the date of the election to the conclusion of the last case found in *Williamson v. Musick,* 59.

17 *Williamson et al.,* 39–41. The case was submitted to the West Virginia State Supreme Court of Appeals on September 29, 1904, and decided on October 18, 1904; *Williamson v. Musick,* 70, 72, 60; In *Little Kingdoms,* Ireland cites examples of the complicity of county court clerks. Robert M. Ireland, *Little Kingdoms: The Counties of Kentucky, 1850–1891* (Lexington: University Press of Kentucky, 1977), 145; Source for Buskirk exclamations: *Williamson v. Musick,* 63–65; Source for temper and humor: William Ely, *The Big Sandy Valley: A History of the People and Country from the Earliest Settlement to the Present Time* (Catlettsburg: Central Methodist Publishing, 1887), 203.

18 *Williamson et al.,* 39–41, and *Williamson v. Musick,* 65–75.

19 At the time of the 1904 election dispute, it should be noted that all five justices of the West Virginia State Supreme Court of Appeals were Republicans: Lewis, *Transforming the Appalachian Countryside,* 112. The role of the Court in solidifying Republican primacy throughout the state is also addressed by Lewis in this work; see Chapters 3 and 8 of *Transforming the Appalachian Countryside.*

20 "United Thacker Coal Company" advertisement, *Williamson Enterprise,* June 4, 1908; Supervisor's report on Mingo County history, January 3, 1938, Historic Records Survey, Mingo County Book I, WVRHC. Lambert served as mine superintendent for several companies: Red Jacket Coal Company (1902), Vulcan Coal Company (1903–1904), Mate Creek Coal Company (1905–1906), Magnolia Coal & Coke Company (1915), and Stone Mountain Coal Corporation (1918); *Annual Reports,* West Virginia Department of Mines, 1902–1918. The largest of Mingo's earliest coal companies, Thacker Coal & Coke, was managed in this early period by T. E. Houston, who ultimately became one of the most powerful coal men in southern West Virginia.

21 *Mingo Republican,* April 20, 1914 and October 28, 1914.

22 "B. Randolph Bias" in 1905 *Progressive West Virginians: Some of the Men*

Who Have Built Up and Developed the State of West Virginia, Robert E. Murphy, compiler (Wheeling, WV: *The Wheeling News*, 1905), 271; "Everett Leftwich" in same, 97; Sources for information on James Little: *Twelfth United States Census* (1900); *Coal Trade Journal* 36 (25 August 1897): 454; *Annual Report of the West Virginia Department of Mines* from 1897–1899; Source for information on S. T. Lambert: 1908 political advertisement from a February 11, 1908, unknown newspaper clipping, Keadle Papers, WVRHC.

23 Confirmation of the assertion that Ferrell, Rutherford, and Hatfield were from pioneering families can be found in the discussion of those families in Waller, Ely, and Jillson; Sheppard's biographical facts can be found in 1905 *Progressive West Virginians*, 101; For explanation of The King case, see Chapter 7 of Cubby, "Transformation of the Tug and Guyandot Valleys."

24 George S. Wallace, *Cabell County Annals and Families* (Richmond: Garrett & Massie, 1935), 482–483; Details of Scherr's movements through the ranks of Mingo's Republican leadership can be found in "Condensed Facts," P8662, Pamphlet Collection, WVRHC.

25 The *Old Liner* appears to have been a brief-run newspaper published during the 1908 political faction wars in Mingo County. It is highly probable that some of the undated, unidentified newspaper articles previously cited from the Keadle Papers came from the *Old Liner*; however, this assumption cannot be verified. Keadle Papers, WVRHC.

26 "To The Honest Voters of Mingo County," 1908 election broadside, Keadle Papers, WVRHC.

27 Source for bolt: *Williamson Enterprise*, June 4, 1908; Source for "Judas": Undated *Old Liner* newspaper clipping, Keadle Papers, WVRHC. "Mont" was M. Z. White and "Jack" was J. K. Anderson, one of Mingo's first county commissioners and the general manager of the United Thacker Coal Company; Source for identities: *Williamson Enterprise*, June 4, 1908, and undated, unknown articles in Keadle Papers, WVRHC; Source for assertion about coal men: *Handbooks of the West Virginia Legislature* (later known as the *West Virginia Bluebook*), 1908–1935, WVRHC.

28 J. R. Booth and S. T. Lambert, 1908 political advertisements in undated, untitled newspaper clippings in Keadle Papers, WVRHC.

29 For an analysis based on group biography, see Kenneth Sullivan, "Coal Men of the Smokeless Coalfields," *West Virginia History* 41 (Winter 1980): 142–

165; Assertion based on a group biography study of the Williamson-Thacker coal elite, compiled for a seminar paper, by the author. Data were drawn from the U.S. Census (1900–1920) and various West Virginia biographical publications, all available in the West Virginia and Regional History Collection.

30   Hatfield correspondence, letter no. 29; "Henry D. Hatfield," *Men of West Virginia*, vol.2 (Chicago: Biographical Publishing Company, 1903): 727–728, 727; Henry D. Hatfield to C. H. Ambler, December 5, 1953, Hatfield Papers, WVRHC; Wallace, *Annals*, 403; Carolyn Karr, "A Political Biography of Henry Hatfield," *West Virginia History* 28 (October 1966/January 1967): 36–37.

31   For information on the Chafin Family in Logan, see *West Virginia Heritage Encyclopedia*, 5: 919–921 and Lawson, 41; Source for Greenway and Henry D. Hatfield controlling Mingo: "Condensed Facts," 7, 1; Source Henry D. "Drewy," William "Bill," and McGinnis "Mac" Hatfield controlling McDowell County: *Sodom and Gomorrah Today, or the History of Keystone, West Virginia*, (n.p.. 1912), no page numbers. "Condensed Facts" is Pamphlet 8662 and "Sodom and Gomorrah" is in the Rare Books Collection, both in the WVRHC; Source for connection between Hatfields and Logan County Chafins: Raymond Chafin, with Topper Sherwood, *Just Good Politics: The Life of Raymond Chafin, Appalachian Boss* (Pittsburgh: University of Pittsburgh Press, 1996), 20; Source for Bill and "Mac" Hatfield's terms as sheriff: McDowell County DAR, *McDowell County* (Fort Worth, TX: University Supply and Equipment Company, 1959), 113.

32   Lee, 137–138; Kenneth R. Bailey, "A Judicious Mixture: Negroes and Immigrants in the West Virginia Mines, 1880–1917," *West Virginia History* 34 (January 1973): 141–161; The report of the *Wiley v. Hughes* case and "Sodom and Gomorrah Today" offer several examples of the white-black political relationships in McDowell County.

33   "Contested Election Case of Rankin *Wiley v. James A. Hughes* from the Fifth Congressional District of West Virginia," Sixty-Second Congress, 2nd session, Report No. 1229. [Mr. Covington, from the Committee on Elections No. 1 submitted the following report to accompany H.Res.703] (Washington: GPO, 1912).

34   "Contested Election Case of *Rankin Wiley v. James A. Hughes* from the Fifth Congressional District of West Virginia," 2–4; Women's Christian Temperance Union of West Virginia, *Report of the Twenty-Sixth Annual Meeting of the Women's Christian Temperance Union of West Virginia, held at Huntington, WV, October 2–6, 1908*, edited by Mrs. K. M. Murill (Charleston, WV: Tribune Printing Company,

1909), 79 and *West Virginia Woman's Christian Temperance Union, Twenty-Eighth Year, Charleston, WV, October 5,6, 7, 1910* (Fairmont, WV: Index Print, n.d.), 113. For more on WCTU activities in West Virginia in this period see: Barbara J. Howe, "West Virginia Women's Organizations, 1880s–1930 or 'Unsexed Termagants' . . . Help the World Along," *West Virginia History* 59 (1990): 81–102.

35   "Contested Election Case of *Rankin Wiley v. James A. Hughes* from the Fifth Congressional District of West Virginia,"2–4; *Williamson Daily News*, August 8, 1914; According to Gary J. Tucker, by 1910, the importance of the African-American electorate in Mingo, McDowell, and several other southern West Virginia counties led to their designation as West Virginia's "Blackbelt" counties; For more on the 1910 election, see: Gary J. Tucker, "William E. Glasscock and the Election of 1910," *West Virginia History* 40 (Spring 1979): 254–267; *Mingo Republican*, November 8, 1912.

36   Steven J. Diner, *A Very Different Age: Americans in the Progressive Era* (New York: Hill and Wang, 1998), 209; John J. Cornwell editorial in *Hampshire Review*, April 11, 1911, quoted in Lucy Lee Fisher, "John J. Cornwell, West Virginia Governor, 1917–1921," *West Virginia History* 24 (April 1963/July 1963): 258–288, 370–389, 272.

37   John J. Cornwell, *A Mountain Trail: From the Farm to Schoolroom, to the Editor's Chair, the Lawyer's Office and the Governorship of West Virginia* (Philadelphia: Dorrance and Company Publishers, 1939), 97.

38   Keith Dix, *What's a Miner to Do: The Mechanization of Coal Mining* (Pittsburgh: University of Pittsburgh Press, 1988), 13.

39   David A. Corbin, *Life, Work, and Rebellion, The Southern West Virginia Miners, 1880–1922* (Urbana: University of Illinois Press, 1981), 6.

40   John H. M. Laslett, *Colliers Across the Sea: A Comparative Study of Class Formation in Scotland and the American Midwest, 1830–1924* (Urbana: University of Illinois Press, 2000), 144–145.

41   Laslett, 144–145; *Annual Reports* 1911, 8; Prior to Mingo's separation from Logan in 1895, production figures for mines in the Thacker coalfield were reported as part of Logan's figures; Paul Salstrom, *Appalachia's Path to Dependency: Rethinking a Region's Economic History, 1730–1940* (Lexington, KY: University Press of Kentucky, 1994), 36, 72–73; Hinrichs, 117.

42   W. P. Tams interview with Richard Hadsell, A&M 2584, typescript pages

3–4, WVRHC; "Shifting the Deal," *Bluefield Daily Telegraph*, February 21, 1895, in David E. Johnston Papers, ERCA.

43   "Shifting the Deal." The southeastern West Virginia coalfields produced a high quality, "low volatile"—hence "smokeless" coal, while the Thacker coalfield produced a lesser quality "high volatile" coal.

44   Andrew Roy, "The Thacker Coal Field of West Virginia," *Mines and Minerals* 19 (May 1899): 472; See also Grace Palladino, *Another Civil War: Labor, Capital, and the State in the Anthracite Regions of Pennsylvania, 1840–1868* (Urbana and Chicago: University of Illinois Press, 1990).

45   August 30, 1898, Bert Wright Diary, Roland Luther Collection, ERCA.

46   Jerry Bruce Thomas, "Coal Country: The Rise of the Southern Smokeless Coal Industry and its Effect on Area Development, 1872–1910," (Ph.D. diss., University of North Carolina, 1971), 237; "Shifting the Deal," *Bluefield Daily Telegraph*, February 21, 1895, in David E. Johnston Collection, ERCA; Laslett, *Colliers*, 144–145.

47   The 1897 strike was called July 4, 1897; in Mingo the strike lasted 30 days. According to the 1898 *Annual Report*, a second attempt at a strike was made on April 1, 1898, hence the reference "the 1897–1898 strike." *1898 Annual Report of the West Virginia Department of Mines*, 47. From 1898–1903, the *Annual Report of the West Virginia Department of Mines* reported yearly on strike activity in the state's coal producing counties; reasons for Mingo's strikes varied from mine to mine. 1898–1900 *Annual Reports of the West Virginia Department of Mines*, Part III: 286. The 1898–1900 *Annual Reports* were published in a single volume, divided by "parts"; Bert Wright Diary, August 30, 1898, Luther Papers, ERCA. Source for willingness of natives to scab: William Carey interview with Keith Dix, October 17, 1971, at Red Jacket, Mingo County, West Virginia, Oral History Collection, WVRHC.

48   Andrew Roy, *History of the Coal Miners of the United States* (Columbus, OH: Press of J. L. Trauger Printing Co. 1902), 395; Source for "injunctions": Roger Fagge, *Power, Culture and Conflict in the Coalfields: West Virginia and South Wales, 1900–1922* (New York: University of Manchester Press, 1996), 114; Corbin, *Life, Work, and Rebellion*, 27. Corbin's failure to acknowledge Mingo's early strike efforts led many to assume that until the events of 1920, Mingo's miners had quiescently accepted their own oppression.

49   *"Thacker Coal & Coke Company v. Burke et al.," Reports of the West Virginia Supreme Court of Appeals* 59 (February 15, 1906–April 24, 1906): 253–262; "Miners in Conference" *United Mine Workers Journal* 11 (4 July 1901): 2; *"Thacker v. Burke,"* 254; 1901 *Annual Report of the West Virginia Department of Mines,* 114; Corbin, *Life, Work, and Rebellion,* 47; S. S. Morrison to John Mitchell, July 3, 1901, John Mitchell Papers, Microfilm Reel 3, Charles C. Wise Library Microfilm Collection, West Virginia University; originals archived at Catholic University, Washington D.C.

50   Jeffreys-Jones, 10, 18; S. S. Morrison to John Mitchell, August 25, 1901, Mitchell Papers; W. H. Crawford to John Mitchell, August 30, 1901, Mitchell Papers; W. H. Crawford to John Mitchell, August 7, 1901, Mitchell Papers.

51   1902 *Annual Report of the West Virginia Department of Mines,* 94; Fagge, 115; C. E. Lively joined the UMWA at Black Band, Kanawha County in 1902: "Testimony of C. E. Lively," February 25–26, 1921, unknown newspaper, Matewan Omnibus Collection, ERCA; *Thacker v. Burke* started in 1901, was submitted to the West Virginia State Supreme Court of Appeals in 1903, and decided in March 1906. How much longer the case lasted is not known, but the Supreme Court's 1906 decision only clarified the case in order to remand it to the circuit court. *Thacker v. Burke,* 261; *The United Mine Workers in West Virginia,* 54; 1902 *Annual Report,* 94; *Huntington Advertiser,* September 5, 1902; West Virginia State Supreme Court of Appeals, *"Thacker Coal & Coke Company v. Burke et al.," Reports of the West Virginia Supreme Court of Appeals* 59 (February 15, 1906 – April 24, 1906): 253–262; Legal document from the *Thacker v. Burke* case, John Mitchell Papers; *Mingo Republican,* July 8, 1920; "Trial Testimony of Jesse P. 'Toney' Webb," Lewis Collection, ERCA.

52   Assertion based on data from the 1910 *Annual Report of the West Virginia Department of Mines,* 8.

53   Corbin, *Life, Work, and Rebellion,* 1; Lawson, 47; Cubby, "Transformation," 173; *Abstract of the Thirteenth Census of the United States, with Supplement for the State of West Virginia* (Washington, D.C.: GPO, 1913).

54   *Abstract of the Thirteenth Census, with Supplement for West Virginia,* 944; Edwin A. Cubby, "Railroad Building and the Rise of the Port of Huntington," *West Virginia History* 32 (October 1970): 246.

55   Ron Chernow, *The House of Morgan: An American Banking Dynasty and*

the Rise of Modern Finance (New York: Simon & Schuster, 1990), 82; Harold W. Houston, Brief on behalf of the United Mine Workers of America, before the committee on Education and Labor, United States Senate, in the matter of the investigation of violence in the coalfields of West Virginia and adjacent territory, (n.p. 1921), 45. Hereafter referred to as "Houston Brief."

56   Corbin, Life, Work, and Rebellion, 4; Joseph T. Lambie, From Mine to Market: The History of Coal Transportation on the Norfolk and Western Railway (New York: New York University Press, 1954), 335; 1922 Annual Report of the West Virginia Department of Mines, 117. Between 1897 and 1904 McDowell County ranked either first or second in production. From 1905 to 1922 McDowell ranked first in the state. In that same period, Mingo ranked from ninth to seventh.

57   Robert Y. Spence, Land of the Guyandot: A History of Logan County (Detroit: Harlo Press, 1976), 289, 318; Lee, 136–137. For more on Chafin rule in Logan see: Lee, 87–121. See also: Spence, Land of the Guyandot, 465–468; The source for "No shoestring operations": Spence, 318.

58   Roy, "Thacker Coal Field": 472; For more on contrast to the N&W's treatment of the Thacker operators, see Spence, 342; Edwin A. Cubby, "Railroad Building," 246. Cubby cites the creation of the Island Creek Railroad Company in 1902 as evidence of Island Creek's financial power and seriousness in its negotiations with the C&O; 1910 Annual Report of the West Virginia Department of Mines, 8. Logan ranked seventh and Mingo ninth out of 34 coal producing counties.

59   Walter R. Thurmond, The Logan Coal Field of West Virginia: A Brief History (Morgantown, WV: West Virginia University Library, 1964), 15.

60   Thurmond, 15–16.

61   Thurmond, 15–16.

62   Thurmond, 15, 22; also, Justus Collins, Justus Collins Papers, WVRHC.

63   Assertion based on data extrapolation from 1902–1904 Annual Reports of the West Virginia Department of Mines; 1904 Annual Report of the West Virginia Department of Mines; Cubby, "Transformation," 246; Source for "crash" and "worrisome regularity": Chernow, 128; Source for "panic": George H. Soule and Vincent P. Carosso, American Economic History (New York: Dryden Press, 1957) 271; Source for "daily": Barkey, 60; Source for "contraction": Ronald Eller, Miners, Millhands, and Mountaineers: Industrialization of the Appalachian South, 1880–1930 (Knoxville: University of Tennessee Press, 1982), 141.

64   Eller, 141.

65   Assertion based on data extrapolation from *Annual Reports of the West Virginia Department of Mines*; Soule and Carosso, 340.

66   Eller, 141.

67   Chernow, 67–69. The corporate consolidation came to be known as "Morganization"; Paul H. Rakes, "Technology in Transition: The Dilemmas of Early Twentieth Century Coal Mining," *Journal of Appalachian Studies* 5 (Spring 1999): 27–60, 32.

68   Curtis Seltzer, *Fire in the Hole: Miners and Managers in the American Coal Industry* (Lexington, KY: University Press of Kentucky, 1985), 13; Keith Dix, *Work Relations in the Coal Industry: The Handloading Era, 1880–1930* (Morgantown, WV: Institute for Labor Studies, Division of Social and Economic Development, Center for Extension and Continuing Education, West Virginia University, 1989), 36.

69   1910 *Annual Report of the West Virginia Department of Mines*; Phil M. Conley, *History of the West Virginia Coal Industry* (Charleston: Education Foundation Inc., 1960), 269.

70   Lacy A. Dillon, *They Died For King Coal* (Winona, MN: Apollo Books Inc., 1985), 39–41. No extant local records exist for this disaster.

71   *Abstract of the Thirteenth Census, with Supplement for West Virginia*, 620–621.

72   Cubby, "Railroad Building," 245–246.

73   Soule and Carosso, 340.

74   *Coal Trade Journal* 36 (February 17, 1897): 899; Thomas, "Coal Country," 237; Corbin, *Life, Work, and Rebellion*, 6; Lambie, 295, 193, 297–298. The southwestern coalfields were Williamson-Thacker, Kanawha and Logan (Guyan); the southeastern coalfields were Pocahontas and New River.

75   Richard M. Hadsell and William E. Coffey, "From Law and Order to Class Warfare: Baldwin-Felts Detectives in the Southern West Virginia Coal Fields," *West Virginia History* 40 (Spring 1979): 275; "'Confidential': Conservative Statement of the Situation of the Coal-Carrying Railroads, in their relation to the Coal and Coke Development of Kentucky, Virginia, West Virginia, and Tennessee (Compiled from an Official Report for 1903)" (n.p.: n.d.). Pamphlet 7080, Pamphlet Collection, WVRHC; Lambie, 41, 295.

76   'Conservative Statement,' P7080, Pamphlet Collection, WVRHC; Lambie, 291, 297–298.

77   Lambie, 297; *Mingo Republican*, March 20, 1911.

78   *Mingo Republican*, March 20, 1911; *History of the Borderland Coal Company*, 9. This document is part of the descriptive materials of the E. L. Stone Papers archived in the Special Collections at the University of Virginia, Charlottesville, Virginia; *The Coal Catalog: Combined with Coal Field Directory for the Year 1920* (Pittsburgh, PA: Keystone Consolidated Publishing Company, 1920).

79   December 3, 1911, newspaper article, probably the *Bluefield Daily Telegraph*, in the Cooper Collection, ERCA; Chernow, 148.

80   *Abstract of the Thirteenth Census, with Supplement for West Virginia*, 620–621, 621, 626; *Coal Trade Journal* 37 (January 5, 1898): 12.

81   *Abstract of the Thirteenth Census, with Supplement for West Virginia*, 592; Hatfield correspondence, letter no. 2.

82   "Smokey" Mose Adkins interview with John Hennen, Summer 1989 Matewan Oral History Project; Corbin, *Life, Work, and Rebellion*, 8; Several Project narrators stressed the independence land ownership afforded their families: Jim Backus, Bertha Damron, Rufus Starr, Addie Nowlin and Stella Pressley; Among the strike leaders not tied to the company system were Charlie Kiser, John Collins, and Fred Burgraff. Sources for this information: (about Kiser): Stella (Kiser) Pressley interview, with Rebecca J. Bailey, Summer 1989 Matewan Oral History Project; (about Collins): Bertha (Collins) Damron, interview with Rebecca J. Bailey, Summer 1989 Matewan Oral History Project; (about Burgraff): Hawthorne Burgraff, interview with John Hennen, Summer 1989 Matewan Oral History Project.

83   Several narrators from the 1989–1990 Matewan Oral History Project discussed the peddling activities of Mingo's farmers: Harry Berman, Edith Boothe, Hawthorne Burgraff, and Daisy Nowlin; Salstrom, "Newer Appalachia as One of America's Last Frontiers," in *Appalachia in the Making*, 92; Salstrom, *Path to Dependency*, 40.

84   1891–1892 *Gazetteer and Business Directory*; 1902–1903 *Gazetteer and Business Directory*.

85   1902–1903 *Gazetteer and Business Directory*. Thacker had 600 residents and Matewan 250; "Condensed Facts," 4, P8662, Pamphlet Collection, WVRHC.

86   Cubby, "Transformation," 246; Collier, 55; Ray V. Hennen, *West Virginia Geological and Economic Survey: Logan and Mingo Counties* (Wheeling: n.p., 1914), 16, 465.

87   Assertion regarding pace of Dingess's decline based on author's interpretation of information offered on "Dingess" in 1902–1903 and 1906–1907 *Gazetteer and Business Directories*; For the story of Dingess and the refinement of political graft and corruption in Mingo County, see Huey Perry, *They'll Cut Off Your Project: A Mingo County Chronicle* (New York: Praeger Publishers, 1972), 49–54; *Williamson Daily News*, August 22, 1918.

88   The impact of industrialization's impact on families has been documented throughout the Appalachian region. See: Sara Lubitsch Tudiver, "Political Economy and Culture in Central Appalachia, 1790–1977" (Ph.D. diss., University of Michigan, 1984), 113.

89   Smith, *Early History of Mingo County*, 6; Hatfield correspondence letter no. 6. The marriage connections between the Hatfields, Ferrells, and Chambers families can be confirmed in: Donna L. Brown, *Logan County Marriages, Book 1: 1872–1892* (Logan, West Virginia: Logan County Genealogical Society, n.d.); E. B. Chambers' purchase of the first lot in Matewan noted in "E. B. Chambers," *Williamson Enterprise*, June 4, 1908; Anderson Ferrell's wives are sometimes identified as Birdie or Bridget (Phoebe Hatfield) and Sarah or Sally Chambers; "Ferrell v. Ferrell," *Reports of the West Virginia Supreme Court of Appeals*, 53 (March 28, 1903–November 21, 1903): 515–524, 516.

90   "Ferrell v. Ferrell," 518–524; Hatfield correspondence, letter no. 6. According to Margaret Hatfield, some Hatfields believed that Matewan had been "stolen" by the Chambers.

91   *Abstract of the Thirteenth Census, with Supplement for West Virginia*, 592. See Catherine McNicol Stock, *Rural Radicals: Righteous Rage in the American Grain* (Ithaca and London: Cornell University Press, 1996), 107; For an Appalachian perspective on the impact of economic crises on violence, see: Lynwood Montell, *Killings: Folk Justice in the Upper South* (Lexington: University Press of Kentucky, 1986), 145 and 160–161.

92   *Williamson Enterprise*, June 4, 1908. The *Enterprise* reported that the men convicted of the attack awaited "additional sentences on additional counts." This issue is one of a set of scattered issues of early Mingo County newspapers; no

other details of the case have been uncovered; Source for link to later violence: James Damron to S. D. Stokes, December 20, 1924, Stokes Papers, WVRHC. The manuscript collections of West Virginia governors John J. Cornwell (1917–1921) and Ephraim F. Morgan (1921–1925) contain several letters critical of the miners' actions during the strike. John J. Cornwell Papers, and Ephraim F. Morgan Papers, WVRHC; Margaret Hatfield interview with Rebecca J. Bailey, Summer 1990 Matewan Oral History Project; Fagge, 214.

93   Again, according to Margaret Hatfield, many residents of Matewan possessed strict views on the rights of property and individual morality. These individuals had little sympathy for the evicted miners, because in their opinion, the company houses were the rightful property of the coal companies. Lynwood Montell observed a similar attitude towards property rights in another Appalachian community. *Killings*, 153; Community ambivalence and/or hostility to labor organization was not unique to Appalachia. In *An Unsettled People*, Rowland Berthoff observed of reactions to Haymarket, Homestead and Pullman that "public sympathy for the underpaid workingman at the outset of a strike was apt to turn, as disorders inevitably followed, to a muddled concern for the property rights of capital." Rowland Berthoff, *An Unsettled People: Social Order and Disorder in American History* (New York: Harper & Row, 1971), 390; In Matewan, alienation from the miners' struggle only intensified in the Massacre's aftermath. In particular, the untimely marriage of Sid Hatfield and Jessie Testerman caused many to believe that perhaps Hatfield had killed Testerman to get Jessie. Hatfield correspondence, letter no. 1 and Harry Berman, interview with John Hennen, Summer 1989 Matewan Oral History Project. It should be noted that Mr. Berman informed Hennen of the "rumors" that swirled around Sid and Jessie rather than imparting a personal opinion.

94   These cases illustrate the survival of traditional attitudes towards, and expectations of, the court and legal system first analyzed by Altina Waller in *Feud*, 47–48. Waller's analysis and therefore the author's, was influenced by, from David Thelen, *Paths of Resistance: Tradition and Dignity in Industrializing Missouri* (Columbia: University of Missouri Press, 1991 [reprint New York: Oxford University Press, 1986]), 77–85.

95   "Hatfield v. Allison," *Reports of the West Virginia Supreme Court of Appeals*, 57 (January 24, 1905–April 25, 1905): 374–384.

96   *"Hatfield v. Allison,"* 374–375, 379, 380–382.

97   *"Hatfield v. Allison,"* 376–377, 378, 382–383.

98   Sources for prominent families: Tudiver, and more recently Dwight Billings and Kathleen Blee, *The Road to Poverty: The Making of Wealth and Hardship in Appalachia* (Cambridge: Cambridge University Press, 2000); Source for contemporary criticism of local juries: Z. T. Vinson, "Railway Corporations and the Juries," *Minutes of the West Virginia Bar Association,* 17th Annual Meeting, (Clarksburg, February 12–13, 1902): 42–51; Sources for visit by Cornwell: Cornwell, *Mountain Trail,* 97; *"Hatfield v. Allison,"* 384, 377; Source for attitude toward contracts: Waller, 47; Source for conflicting philosophy of the judges: Lewis, *Transforming the Appalachian Countryside,* Chapter Four, especially page 110.

99   *"State of West Virginia v. Elias Hatfield,"* in *Reports of the West Virginia Supreme Court of Appeals,* 48 (April 14, 1900–December 21, 1900): 561–576, 571–2; Otis K. Rice, *Hatfields and McCoys* (Lexington, KY: University of Kentucky Press, 1982), 120.

100   *Logan Banner,* July 6, 1899, quoted in Cubby, "Transformation," 178; *Logan Banner,* July 13, 1898, quoted in Cubby, "Transformation," 178; Rice, *Hatfields and McCoys,* 120; Virgil Carrington Jones, *The Hatfields and the McCoys* (NY: Ballantine Books, 1948), 216–217 and 226–227; Kenneth R. Bailey, "A Temptation to Lawlessness: Peonage in West Virginia, 1903–1905," *West Virginia History* 50 (1991): 25–45, 37–38.

101   *"State v. Hatfield,"* 572, 575; J. C. McWhorter, "Abolish the Jury," *West Virginia Law Quarterly* 29 (January 1923): 97–108. Based on available biographical information on the McWhorter family, it is unlikely that the two McWhorters were closely related. See *West Virginia Heritage Encyclopedia,* volume 14.

102   Edwin S. Doolittle, "On the Circuit in Southern West Virginia," *Greenbag* 12 (1900): 284–286, 285; *Bluefield Daily Telegraph,* 20 March 1921, *Bluefield Daily Telegraph* Collection, ERCA; "State v. Hatfield," 563.

103   *Bluefield Daily Telegraph,* March 20, 1921, *Bluefield Daily Telegraph* Collection, ERCA; *"State v. Hatfield,"* 563; *New York Times,* March 21, 1921.

104   In addition to the 1911 incident, in 1914, another of Matewan's police chiefs, O. L. Ackerman, died in a "duel to the death" with the mayor's nephew. *Mingo Republican,* February 13, 1914.

105   *Williamson Enterprise,* April 27, 1911; and *Mingo Republican,* April 28, 1911.

106   *Mingo Republican,* April 28, 1911.

107   *Williamson Enterprise,* April 27, 1911; *Williamson Enterprise,* 4 May 1911.

108   *Mingo Republican,* October 20, 1911; "Thomas Chaffin (Chafin)," *West Virginia Heritage Encyclopedia,* 5: 919.

109   *Mingo Republican,* October 20, 1911.

110   *Williamson Enterprise,* April 27, 1911; *Williamson Enterprise,* May 4, 1911.

111   Ernie Reynolds, "After the Round-Up Where Do all the Brave Men Go?," Ernie Reynolds Collection, ERCA.

2

THE PROGRESSIVE ERA?

"Political strife seems to be perennial in this county."
– *Williamson Daily News, February 1917*

MINGO COUNTY POLITICS between 1912 and 1919 followed the trend that had begun in 1895. Every election was bitterly contested with charges and countercharges of graft, illegal voting, and wholesale election fraud and theft. In West Virginia during the Progressive Era, reform party movements were led more often by "disgruntled outs trying to get back in" than by genuine reformers. This meant that in Mingo County, the appearance of third-party tickets such as Progressive, Prohibition, and Bull Moose on election ballots merely re-labeled old factions and party splinter groups. An important yet minor exception was the appearance of the Socialist party in the 1912 election. However, primary and ballot "reform" legislation minimized the impact of the Socialist party locally and around the state. County and municipal elections from 1912 through 1919 in Mingo remained extremely fractious contests between the two main parties, the Republicans and Democrats.[1]

There was one significant difference between the elections of the 1912–1919 period and previous elections—the pendulum of political power swung farther with each election. The period began with the Hatfield machine in control of Mingo County. However, their abuse of power and flagrant manipulation of the political process led to the Hatfields' eclipse in 1916. Temporarily united in opposition to the Republican Hatfields, Mingo's Democratic Party regained control of the county in 1916, only to resume its internecine struggles shortly thereafter. Neither Progressive reform agendas nor the Great War effort inhibited the ongoing faction wars of Mingo County politics. When UMWA organizers arrived in Mingo County in the spring of 1920, their early success depended, in part, on their ability to capitalize on the volatility of the local political scene.

The 1912 general election marked the Hatfield family's triumphant political ascendancy in West Virginia. Native son Henry D. Hatfield, who had migrated to McDowell County and joined the coal operator-backed Republican Party, combined a traditional politicking style, a Progressive reform platform, and the coercive strategies of a machine boss, to win election to the governor's office. Riding his coattails, Henry D.'s older brother Greenway led the Republican sweep of Mingo's county offices of sheriff, circuit court judge, prosecuting attorney, assessor, county commissioner, and state delegate. Between 1912 and 1915, Henry D. and Greenway Hatfield shrewdly co-opted the language of reform to strengthen their control over state and county politics. At the same time, the Hatfield brothers' bald abuse of power fractured Henry D.'s syncretistic powerbase and inspired a united front of Hatfield opponents who toppled the Republicans from power in 1916.[2]

Henry D. Hatfield oversaw the enactment of more reform legislation than any other West Virginia governor in the Progressive Era. During his tenure as the president of the West Virginia senate and his four-year term as governor, Hatfield championed primary and ballot reform, prohibition, bills that called for the eradication of the much hated mine guard system, the establishment of West Virginia's workmen's compensation fund, and Public Service Commission. Hatfield's success in achieving the passage of these measures emerged largely from his unique political style, which allowed him to elicit support from all segments of West Virginia society, from rural and urban areas and from the working class to the industrial elite.[3]

Running for the governor's office in the midst of the Paint Creek and Cabin Creek Strike of 1912–1913, Hatfield, who hailed from McDowell, West Virginia's most operator-controlled county, won the support of West Virginia's laboring classes by publicly committing himself to addressing the issues that mattered most to them, including strengthening West Virginia's laws against the use of private guards. One incident that revealed how Hatfield convinced miners of his sincerity also exposes his canny use of traditional mountain politicking. During a campaign stop in Williamson, Hatfield spotted in the crowd A. D. Lavinder, a childhood friend who had become one of southern West Virginia's leading Socialist miner activists.

Lavinder later recalled that after concluding his speech, Hatfield called him aside and told him that together they "could run these mine guards out of here." True to his word, once in office Hatfield shepherded the Wertz mine guard bill through the West Virginia legislature. However, the absence of a key component of the bill illustrates how Hatfield eluded censure from the coal elite—the Wertz bill had no enforcement clause.[4]

As in the case of the Wertz bill, Hatfield's other reforms contained clauses that diffused their threat to the industrial elite. For example, the 1915 primary election bill required voters to register their party affiliations and in turn opened precinct registers to public display. This "reform" measure limited the free and confidential exercise of the franchise, especially in company towns or even "independent" towns where local coal company officials frequently were appointed poll registrars or worked as poll clerks. Thanks to the 1915 bill, these officials were now privy to the political affiliations of their employees. The 1915 primary bill also required would-be candidates to pay a registration fee to enter a primary election. Reflective of the popularly held Progressive belief that political candidacy and public offices should be restricted to people who were supposed to be more capable of acting in the public interest, the fee requirement did not impede the influence of machine politicians. The "reform," intended to limit political candidacy to the "better classes," merely raised the stakes and honed the machines' desire to select absolutely loyal and controllable candidates.[5]

With little interest in coalition building or bipartisan politics, Hatfield also did not shrink from cowing antagonists with physical violence. Before his election as governor, Hatfield pummeled a racist newspaper editor who had criticized his political reliance on African-American votes. Even after becoming governor, Hatfield resorted to using his fists. While trying to resolve the Paint Creek strike, Hatfield allegedly knocked a recalcitrant coal operator into a corner of his office. As another example, when a Public Service Commissioner (his own brother-in-law) threatened to go against the governor's wishes in a well-publicized utility case, Hatfield invited the man to a meeting in the governor's office. As the unsuspecting victim seated himself, Hatfield leaped upon him and beat him nearly unconscious.[6]

Although Greenway Hatfield was less openly and personally confrontational than his younger brother, his rule of Mingo County from 1912–1916 exposed the machine component of the brothers' power base. Sheriff Hatfield created an "army of deputies" staffed by an array of cousins and hangers-on. Political subordinates paid cash or delivered votes in return for appointment to public office. During his tenure as sheriff, Greenway Hatfield used inmates from the county jail to cultivate his farm. He then used the proceeds to feed them and pocketed the money he saved from the county's prisoner maintenance fund. Also, as one Hatfield descendant noted, it was not uncommon to find Greenway's enemies floating in the Tug Fork River.[7]

Only one thing kept the Hatfield machine from ruling Mingo County absolutely. Control of Williamson, the county seat, had always lodged in the hands of the Democratic relatives and business associates of the city's founder Wallace J. Williamson. By 1915, when Henry D. Hatfield decided to use his control of the West Virginia legislature to break the Democrats' hold on Williamson, the mayor, A. C. Pinson, had served seven consecutive terms. The story of the revocation of Williamson's municipal charter and the installation of commission government stands as the ultimate example of the Hatfield machine's cooptation of Progressive reform methods for base political ends.

Although he later claimed that "the damned thing made him want to puke" every time he thought of it, Mingo's Republican state senator Wells Goodykoontz proposed a bill to revoke Williamson's charter in the spring session of the 1915 West Virginia legislature. The bill's passage removed seven-term Democratic mayor A. C. Pinson and the city council and replaced them with a bipartisan commission. The bill, which passed on a strictly partisan vote, called for Williamson to be governed from July 1, 1915 until July 1917 by an appointed five-member bipartisan commission.[8]

Rescinding the charters of machine-controlled, corrupt municipal governments and replacing them with bipartisan commissions had been a popular Progressive reform across America since it was most famously applied in Galveston, Texas, in 1901. In the case of Williamson, the Democratic "City Ring" as it had been dubbed, was not a stereotypical Progressive Era urban-ethnic machine. Its leadership was southern white, Anglo-Saxon Protestants, including prominent native businessmen such as Wallace J. Williamson and college-educated migrants such as S. D. Stokes. The Williamson machine manipulated the lower-class industrial and railroad workers who lived within the city limits but did so as an agent of corporate capitalism; the upper echelons of the machine included men like Stokes, who was an attorney who brokered land and mineral deals.[9]

The *Williamson Daily News* declared the Charter Bill a legislative fiat concocted by Governor Hatfield and his cronies to steal for the Mingo Republicans what they could not win in an honest election. As said before, the bill called for the ouster of the Democratic mayor and council and their replacement with appointed commissioners who would serve two years before a new municipal election would be held. Among the new commissioners was the editor of Mingo's Republican newspaper, Orland H. "O. H." Booten, whom the *Daily News* renounced as a "liar, thief . . . degenerate . . . [and] drunkard who [could] be hired to desecrate a graveyard if it were necessary for his own gain."[10]

The *Daily News* began referring to Governor Hatfield as the "Emperor of Williamson," whose lust for absolute power posed a threat to any elected municipal officer in the state who disagreed with or offended him. Hatfield's ability to strip the Williamson Democrats of their control of Mingo's county seat evoked a vicious screed from the *Daily News* regarding the source and nature of his power: "It is true that the city elections are not dominated by negroes, dagoes, hunks, criminals and mercenaries as the county is, but under this law it will be."[11]

When the deadline arrived for the mayor and the other municipal officers to pass control of Williamson over to the appointed commissioners, Mayor Pinson and City Clerk John S. Hall refused to surrender their offices. Both sides filed lawsuits, and despite his periodic defiance of the Mingo Republican machine and his Hatfield cousins, Circuit Court Judge James Damron upheld the legality of the Charter Bill. Ousted Democrats appealed Damron's decision to the West Virginia Supreme Court of Appeals on December 12, 1915.[12]

Five separate Williamson Charter cases were submitted to the high court, but the judges decided that because they all dealt with the same legal issues, a single decision sufficed to explain the court's ruling. With the exception of the president of the court, Justice Poffenbarger, the judges agreed that the central issue of the controversy was whether the supremacy of a state legislature superseded the right of municipal home rule. Despite Justice Poffenbarger's impassioned plea that the letter of the law was protecting corruption, the majority decision upheld the lower court's ruling, and the new Williamson charter went into effect on December 17, 1915. Mayor Pinson and the old city council "quietly . . . handed over the reins of government" on January 1, 1916.[13]

The dislodging of the Democrats from control of Williamson should have ushered in an era of Republican primacy at least locally, but two fatal internal weaknesses had already precipitated the decline of the Hatfield machine, both in Mingo and at the state level. First, the Hatfield brothers' unabashed reliance on corrupt and coercive election strategies had repeatedly embroiled them in political scandals that tainted Henry D.'s reputation as a reformer. Second, once in the governor's office, Henry D. fractured his alliance with the coal elite, first by ignoring his debts to former patrons and later by openly defying their wishes. By 1916, the last year of Henry D.'s gubernatorial tenure, the accumulated antipathy generated by these two issues led to a Democratic sweep of the governor's office and Mingo County's political offices.

The beginning of the end of the Hatfield brothers' reign actually occurred during the first legislative session of Henry D.'s governorship and illuminated both of the Hatfields' major flaws. In 1913 a bribery scandal not only brought national notoriety to the state, it portended the Hatfields' eclipse for two reasons. First, it was only one in a series of political incidents involving the Hatfields and/or their associates that seemed to validate their opponents' charges of corruption. Second, the issue at the heart of the scandal, the selection of a United States senator, opened the schism between Henry D. Hatfield and West Virginia's coal and industrial elite.[14]

When the state representatives converged on Charleston in early 1913, one of the critical issues facing the West Virginia legislature and incoming governor Henry D. Hatfield was the selection of a United States senator. After Stephen B. Elkins' death in 1911, Governor William Glasscock appointed Elkins' son Davis as a temporary replacement. When the legislature convened in 1913, the younger Elkins expected to be granted the senatorship in his own right, but he faced stiff competition from two of southern West Virginia's leading industrial elites, Isaac T. Mann and William S. Edwards. Elkins' "reactionary" views and assumption that the senatorship was automatically his angered Glasscock, and his candidacy was not seriously considered. On the other hand, Mann had contributed generously to the Republican war chest for years. Edwards had also been a longtime loyal party promoter. Both men felt they had earned the honor of representing West Virginia in the United States Senate.[15]

Governor-elect Hatfield's reluctance to back Mann, his old patron in McDowell County politics, forced Mann to compete with Edwards for support among the legislators who would select the next senator. W. P. Tams, another prominent southern West Virginia coal operator, described the events that created the resulting scandal:

> Til then [1913] it was a reasonably priced legislature . . . But unfortunately for Mann another West Virginian, named Edwards . . . wanted to be senator. So they began bidding against each other . . . the thing got so scandalous that more sensible people woke up and told both Mann and Edwards that they would have to withdraw and let someone else go up. Even West Virginia couldn't handle a scandal like that.

After twenty-three ballots, the legislature selected neither Mann nor Edwards. State Senator M. Z. White of Mingo County then threw his support to a third candidate, Nathan Goff, who won the honor. Hatfield rewarded White by appointing him warden of the state penitentiary. When Hatfield and the legislature selected Nathan Goff to be the new senator, Mann withdrew to southern West Virginia and nursed his resentment against Hatfield, whom many felt had betrayed his old political benefactor.[16]

William D. Ord, superintendent of the Red Jacket Consolidated Coal & Coke Company in Mingo County, wrote of Hatfield's behavior to Judge John W. Mason, "Hatfield's personal and political ambitions have become so great that he has . . . turned his back upon some of his friends, especially those in the coal industry, and has . . . actually 'arrayed' the masses against the classes." The final straw for the industrial powerbrokers came when Hatfield chose former State Supreme Court Judge Ira Robinson to succeed him as the 1916 Republican nominee for governor.[17]

There were at least two reasons for the Republicans' defiance of the governor. In 1913, during the Paint and Cabin Creek strike, Judge Robinson had broken with the court to criticize the use of martial law restrictions against the miners. T. E. Houston, president of Thacker Coal & Coke Company, and many other anti-union operators of southern West Virginia, viewed Robinson as a pro-labor traitor to the party and organized "a fight against the Hatfield gang" which was Robinson's main support. Amid such discontent, Isaac T. Mann and Davis Elkins, the disappointed candidates from the 1913 bribery scandal, allegedly forged an alliance to defeat Hatfield. The anti-Hatfield forces chose Secretary of State A. A. "Abe" Lilly of Raleigh County to challenge Judge Robinson for the Republican gubernatorial nomination.[18]

The severity of the discord among the Republicans even divided the party in Hatfield's home county during the 1916 Robinson-Lilly primary race. The Mingo Republicans faced a terrible dilemma. The Hatfield brothers, though still powerful, were estranged from the business wing of the party. On the other hand, the coal operators' candidate, Abe Lilly, had publicly opined that the Williamson charter bill should be overturned, an act that threatened to return the Democratic City Ring to power in Williamson.[19]

When Governor Hatfield and Ira Robinson stopped at Matewan and Red Jacket during their campaign tour of southern West Virginia, Henry D. Hatfield could no longer ignore the hostility his actions had caused. On May 27, 1916, Matewan received its hometown boy bedecked in pictures of Abe Lilly. When Hatfield found a large Lilly banner displayed in the lobby of the Urias Hotel, he refused to eat there and ate at a lunch counter on the street. As the *Williamson Daily News* observed, when Hatfield spoke to a crowd of "pistol-toting deputies and 30 negroes," his famous temper erupted. He "denounced . . . Lilly as an unworthy degenerate . . . [attacked] William D. Ord and boasted that no coal operator could control [him]." Hatfield also vilified Matewan for insulting him by displaying Lilly's picture.[20]

The atmosphere proved no better for Hatfield at Red Jacket. Although he spoke at the carnival grounds, the crowd was small. Young men in automobiles drove back and forth on the road opposite the gathering and with "jesting and singing, so disturbed the governor that he had to pause in his remarks." The open hostility at Matewan and Red Jacket so disconcerted Robinson, he refused to disembark at Williamson. The *Daily News* hypothesized that Governor Hatfield's behavior had "made . . . many bitter enemies" who opposed candidate Robinson.[21]

Despite Mingo's unenthusiastic welcome for Hatfield and Robinson, the Hatfield-Robinson ticket defeated the Damron-Lilly faction in Mingo's primary. Hatfield-Robinson candidates for local and state offices won with majorities of 600 to 1,100 votes. In McDowell County, the Hatfield-Robinson faction also triumphed in the primary, but the two counties were out of step with southern West Virginia's Republican majority. Lilly carried the southern West Virginia Republican counties that were not Hatfield-machine dominated.[22]

While in Chicago for the national convention, Hatfield further exacerbated the southern West Virginia operators' animus toward him. While engaged in an argument with fellow conventioneer and former patron Isaac T. Mann, Hatfield "sucker punched" him after their mutual friends stepped between the two men, according to Mann. Responding to a sympathy letter from Justus Collins, who referred to the governor as "that ruffian Hatfield," Mann denied being injured seriously and noted that Hatfield retreated quickly, "as all bullies and cowards are inclined to do."[23]

After Robinson defeated Lilly in the primary, Lilly did not concede defeat until July 27, 1916. Although their hand-picked candidate lost, West Virginia's Republican coal operators simply regrouped and, via the West Virginia Coal Operators' Association, swung their financial support to Democratic gubernatorial candidate John J. Cornwell. Historically, this move has been ascribed to the operators' desire to have a man who was on their side in the governor's office. However, a letter from Williamson Democrat S. D. Stokes to U.S. solicitor general and West Virginia native John W. Davis adds a new dimension to the operators' actions. As Stokes informed Davis, the friends of General Lilly turned to Cornwell, the Democrat, because they believed that "to ever get as much as a 'toehold' in the halls of the G.O.P. they will have to defeat the Hatfield crowd . . . for they will never receive quarter at the hands of the Hatfields." The operators hoped Robinson's defeat would loosen Hatfield's stranglehold on southern West Virginia's Republican organization.[24]

1916 brought an end to the era of the Hatfield brothers' rule of West Virginia and Mingo County. Governor Hatfield's autocratic leadership of the party and defiance of the state's industrial elite bolstered the Democratic Party's initiative to regain control of the political arena. Sheriff Greenway Hatfield's flagrant subversion of democratic processes in Mingo not only renewed the county's infamy, it also advanced Democratic attempts to return to power at the local level.[25]

To understand why the decline of the Hatfields was so instrumental to the "Democracy's" resurgence, one must examine what had kept the Democrats from challenging the Hatfields' dominion before 1916. The Democratic Party in West Virginia had been a coalition party that, since the dark days of the early 1890s, had been unable to regain control of the state government from the Republicans, who had forged a "party-army" under the guidance of Nathan Goff and Stephen B. Elkins. Divided between Bryanites and competing industrial powerbrokers, the Democratic Party leaders consistently failed to capitalize on the increasing success of Democratic candidates for state legislative seats. However, the death of Stephen B. Elkins three days into the 1911 legislative session presented the Democrats with a unique opportunity. In control of the House of Delegates, the Democrats used the even split in the number of state senators to force the selection of two Democrats for West Virginia's United States senators. Seeking unity, the Democrats appointed the party's most contentious faction leaders, Clarence W. Watson of Fairmont and William E. Chilton of Charleston. Because they both aligned with the pro-industrial wing of the party, Chilton and Watson formed a united front to defeat the liberal faction of the party. However, as illustrated by the faction wars of the Democrats in Mingo County between 1912 and 1914, the loyal subordinates of Chilton and Watson continued to struggle among themselves long after the two principals had found common ground.[26]

The Mingo Democrats' alliance with state level Democratic leaders was first revealed during the 1912 election. In that campaign the *Mingo Republican* accused the Williamson Democratic ring of having at the last minute "adopted underhand tactics in the interest of Boss Watson," for which they were rewarded with cash. "Boss Watson" is a reference to the faction-leader, coal operator, and United States Senator Clarence Watson. However, the Republicans never raised any specific reason or issue that linked the Mingo County Democratic Party with Senator Watson; they only implied that the Democrats had taken Watson's money and had thereby become his hirelings.[27]

Two years later, the 1914 campaign exposed the local significance of the Mingo Democratic alliance with Senator Watson. At a meeting of the Mingo County Democratic Executive Committee in the spring of 1914, G. T. Blankenship of Matewan stepped down from his position on the committee. During the debate that followed, a faction led by Wallace J. Williamson "forced" Executive Committee Chairman W. N. Cole to resign and voted to replace him with Williamson. Known locally as "the Old War Horse," Wallace J. Williamson was the leader of the Williamson Democrats, pejoratively known as the "City Ring." As summer faded into fall, the purge of Blankenship and Cole resurrected the old Williamson-Matewan rivalry. Table 7 reveals the identities of the leading members of the two major Democratic factions, several of whom would play critical roles in the events of 1920–1922.[28]

"Rural" or "County" Democrats (pro-Chilton)	"City Ring" (pro-Watson)
Leader: W. N. Cole	Leader: Wallace J. Williamson
E. B. Chambers (and family)	G. R. C. Wiles
G. T. Blankenship	A. C. Pinson
Dr. R. G. White	W. A. Hurst
J. H. Green	S. D. Stokes
W. F. Hutchinson	
Boyd Adkins	

Table 7. Democratic Factions in Mingo County, 1912–1916
Source: "Mingo Republican," May 21, 1914, and August 6, 1914.

In an anonymous letter published in the *Mingo Republican*, a "Matewan Democrat" listed past and current grievances held by the Matewan and County Democrats against the Williamson gang. Despite the earlier betrayals of these City Ring Democrats, Boss Williamson trusted them over loyal Democrats elsewhere in the county. Rather than select a Matewan Democrat to work the precinct polling, the Old War Horse sent Williamson lawyer and party hack Alex Bishop to Matewan to serve as registrar. The single largest group of voters was the county's coal miners who lived primarily in two districts: Lee, which had Williamson as its heart, and Magnolia, whose center was Matewan. The City Ring controlled the Democrat-dominated Lee district, but Matewan and the votes of the miners of the Magnolia district traditionally belonged to the Republicans. The Old War Horse did not trust the Matewan Democrats to wrest the votes of Matewan and Magnolia District from the Republicans. Within a week of his arrival in Matewan, Bishop faced bribery charges.[29]

Despite the Democrats' desperate accusations and mercenary appeals, the Republicans again swept the Mingo County election in 1914. Republicans were elected state senator, delegate, county school superintendent, county commissioner, county court clerk, and circuit court clerk. Two years later, a report that during the 1914 election "two armed and masked men entered the little school house which served as a polling place" for the Rockhouse precinct and stole the ballot box sparked a political and legal scandal that lasted until 1918. All of the accused men were Republicans, including County Chairman Greenway Hatfield. Although tried in federal court, their prosecutors were southern West Virginia Democrats, one of whom, D. E. French, was a personal enemy of Governor Hatfield. The resuscitation of the 1914 election fraud cases stemmed from two critical political developments: the revocation of Williamson's city charter in 1915 and the bitter partisanship of the local, state, and federal general election campaign of 1916.[30]

Having lost control of Williamson's city government in 1915 and being unable to recapture its position until the next municipal election in 1917, Boss Williamson's City Ring machine lost ground to the County Democrats led by the Chambers' family of Matewan. In the interim, two developments affected the "City Ring's" efforts to remain a viable force in county politics. Southern West Virginia's Republican elite openly broke with Governor Hatfield in the spring of 1916 over the issue of the Republican nomination for governor. Anxious to capitalize on the Republican schism, southern West Virginia Democrats embraced the dissidents in the spring and summer of 1916.

One prominent Mingo County anti-Hatfield Republican was Hatfield cousin and Mingo County Circuit Court Judge James Damron. In 1916, Damron declared that his primary objective was to clean up the voter registration rolls by "striking from the list . . . illegal voters, dead men, mules and tombstones." Damron's other goal was to stop "Bob Simpkins from stuffing . . . ballot boxes," a direct reference to the irregularities in the 1914 election allegedly orchestrated by his Hatfield cousins. Greenway Hatfield and Wells Goodykoontz dismissed Damron's attempts to "clean" the voter registration rolls by claiming that by Damron's method "peace will be converted into riot and chaos will take the place of law and order." Greenway Hatfield's prediction of violence did not dissuade Damron. The State Supreme Court granted Damron a peremptory writ, but nothing else was heard of the issue until May 17, 1920, when clarification of Mingo's voter registration rolls resurfaced.[31]

Two articles in the *Williamson Daily News* disclosed the local reasons for the Republican defection in Mingo in 1916. Yet again, the complaints also reveal how traditional mountain political tactics had survived into the industrial era. On June 2, the *Daily News* asserted that Greenway Hatfield had appointed "pardoned criminals and deputy sheriffs . . . to conduct the primary elections." The *News* also rhetorically asked "how many brothers, cousins, uncles, and sons-in-law of candidates were appointed on the election board?" In case anyone questioned the veracity of its accusations, on June 5 the *Williamson Daily News* published the names and "offenses" of the precinct election officers appointed by Greenway Hatfield. Table 8, while omitting the names of all the officers except one, presents the nature and scope of Hatfield's network. As the table illustrates, all of the 1916 Republican precinct officers were Hatfield relatives, owed their freedom from jail to the Hatfields, or relied on serving Sheriff Hatfield for their income.[32]

The charges of the Hatfields' corruption attracted national attention. According to Toney Webb, the correspondent-author of the *Matewan News*, "the citizens of Matewan have borne the brunt of political dishonesty until we are being looked upon by citizens elsewhere as possessing horns like a rhinoceros and . . . with bristles on our backs like the wild boars of Africa." Webb's fears regarding outsiders' impressions of Matewan and Mingo County were inspired by the comments of a federal official who had been sent to investigate election conditions in Mingo County. Described by the *Daily News* only as "a man high up in the secret service of the United States government," the officer observed to the newspaper's

Precinct	Offenses and/or Familial or Employment Connections of Republican Primary Precinct Officers and Greenway Hatfield
Dingess	1- indicted for graft of school bonds 1- ex-convict and deputy sheriff
Dempsey	1- indicted 4 cases of graft of road funds, deputy sheriff
Naugatuck	2- deputy sheriffs
Rockhouse	1- pardoned criminal, county "sealer of weights" 1- father is candidate for assessor
East Williamson	1- brother is house of delegates candidate 1- brother-in-law is candidate for sheriff
West Williamson	1- deputy sheriff and county road engineer 1- deputy sheriff and prohibition officer
Upper Red Jacket	3- all deputy sheriffs and Hatfields or Hatfield in-laws
Lower Red Jacket (Matewan)	2- both deputy sheriffs and had been charged with padding the voting rolls with 117 names, pardoned by Governor Hatfield
Magnolia	1- notorious for selling whiskey 1- Hatfield first cousin and deputy sheriff
Thacker	2- not residents of West Virginia, work for coal company of J. K. Anderson who was appointed to Public Service Commission by Governor Hatfield
Lick Fork	1- pardoned convict 1- deputy sheriff 1- not resident of precinct- "imported"
Devon	2- deputy sheriffs and first cousins of R. L. "Bob" Simpkins who was also a cousin of Henry D. and Greenway Hatfield
Varney	2- son and foster son of deputy sheriff
Glenalum	1- pardoned criminal
War Eagle	Bob Simpkins (defendant in the 1914 election fraud case)

Table 8. 1916 Republican Primary Election Precinct Officers Appointed by Greenway Hatfield *Source: "Williamson Daily News," June 5, 1916*

correspondent "that Mingo County was the most corrupt politically of any county in the country."[33]

The federal investigator sent to Mingo was only one member of an entire contingent that scoured southern West Virginia in the summer of 1916, searching for evidence of election fraud. Charges were filed against Republican politicians in McDowell and Mingo Counties, including Colonel Edward O'Toole of the U.S. Steel coal mines (McDowell) and Sheriff Greenway Hatfield (Mingo). The federal government dropped charges against Colonel O'Toole but prosecuted the Mingo politicians in two separate cases, referred to as the "War Eagle Case" and the "Rockhouse Case" after the precincts where fraud was allegedly committed.[34]

Before the cases went to court, Mingo County's newspapers fought over why charges were brought against several of the county's Republicans. The *Williamson Daily News* claimed that it was to stop "Wholesale Election Rascality." The *Mingo Republican* reprinted an editorial from the *Huntington Herald Dispatch*, which stated that "the sole aim" of these cases was "to injure the Republican Party and incidentally those leaders in the forefront of the fight against the Chilton-Watson wing of the Democracy." The *Dispatch* offered six reasons why it believed that the cases were politically motivated. The two most important reasons were that the federal officials "only investigated alleged Republican fraud cases" and "the employment of a partisan Democratic lawyer and office seeker as special prosecutor shows that the Federal political machine was not above handing a sop to a faithful ally who had sought much and received little." Despite the outcry of southern West Virginia's Republicans, both the War Eagle Case and the Rockhouse Case proceeded in the fall of 1916.[35]

The *United States v. R. L. Simpkins, et al.* case, also known as the "War Eagle Case," convened in federal district court in Huntington in late September 1916. The prosecution presented a strong circumstantial case against the Mingo Republicans. A string of witnesses revealed how the Hatfield machine bribed, bullied, and carelessly exhibited its power during elections. Elihu Boggs, a coal miner who had been elected clerk of the county court on the Republican ticket, revealed the price of Hatfield patronage. During "several conversations with Greenway Hatfield," Boggs was told he would have to contribute a thousand dollars to the campaign. When the ballots were removed from the boxes, and all but 11 of the Democratic ballots were burned. Only a total of 65 votes had been cast, but replacements were created to make the vote total equal to the number of registered voters in the precinct. Two women teachers provided the final and most extreme example of the Hatfield machine's arrogant display of power. Miss Blanche Depew said "Simpkins told her a mule had voted and that this was more than she could do." Miss Gertrude Rader corroborated Depew's story, adding the detail that "the mule had been voted as 'Vicie Coffee.'" The prosecution intended the stories of Simpkins' insensitivity and crass treatment of the two women to underscore his base moral character and, by association, implicate the Hatfields as abusers of women.[36]

THE COAL DIGGER CANDIDATE

For Clerk of the County Court Vote For
ELIHU BOGGS
17 Years in the Mines of Mingo County

Elihu Boggs, 1914. *Elihu Boggs, a coal miner, served terms as mayor of Matewan and in 1920 was clerk of the Mingo County Court. [Mingo Republican, 1914, WVRHC]*

The trial of the War Eagle election fraud case ended on October 5, 1916, after just one week. Judge Woods instructed the jury to find Greenway Hatfield not guilty, and after a 30-minute deliberation his co-defendants were also exonerated. The defendants had no time to rejoice, however, because the court turned immediately to the Rockhouse Case and impaneled a jury. The primary defendants in the Rockhouse and War Eagle cases were the same.[37]

As in the War Eagle Case, the Rockhouse Case focused on the alleged attempt by members of Greenway Hatfield's machine to steal the 1914 election. However, the Rockhouse controversy centered on the actual theft of the precinct's ballot box by two armed and masked men. The difficulty faced by the prosecution in this case centered on the attempt to expand the case beyond the prosecution of the thieves into a conspiracy trial. Judge Woods' instructions to the jury included a review of election laws and a reminder that "the greatest menace to the purity of election was the power of unlawful combinations to control the voters or nullify their votes." Judge Woods' definition of conspiracy influenced the outcome of the trial. He instructed the jury that "a combination to persuade men to vote is not a conspiracy, but a combination of two or more persons is a conspiracy if it is formed to intimidate, threaten, oppress or injure one or more in a certain way or not to vote at all." After a day's deliberation, the jury remained deadlocked, leaving Judge Woods no option but to declare a mistrial.[38]

In the midst of the War Eagle and Rockhouse cases, attention in Mingo County turned to the progress of the general election campaign. Although the War Eagle and Rockhouse Case did not end in convictions, they upset the Mingo Republicans' equilibrium. Throughout the month of October until the eve of the election, the Republicans maintained a constant barrage aimed at the Democrats, as though they and not the Democrats were trying to regain control over the county. By contrast, the Mingo Democrats confined themselves to their traditional tactic—race-baiting.

The Republicans also tried to maintain the loyalty of the miners, the county's single largest voting bloc. The *Mingo Republican* asked, "Has any labouring man had an increase in wages in the last year commensurate with the increase in the cost of living?" The newspaper then reminded its readers who was responsible for this state of affairs by pointing out that L. E. Armentrout and Harry G. Wilburn, two Democratic candidates for the county court, were "a successful coal operator and . . . an employee of a big coal corporation." During a public speech in Williamson just days before the election, Governor Hatfield himself made the final plea for the miners' vote. After reading a list of local beneficiaries of the workmen's compensation law, Hatfield pointed out that it was "a measure threatened with destruction if Boss Watson succeeded in electing his man Cornwell."[39]

Despite the no-holds-barred effort of the Republicans, the Democrats

swept the 1916 general election in Mingo County. There were several reasons for the Republican failure. First, in 1916, the County Democrats, not
the "City Ring," dominated the slate of candidates. Second, the Republicans
had criticized county commission candidates Armentrout and Wilburn
for "practically [being] strangers to the citizens of Mingo County." This
had proven to be another ill-fated ploy because the Chambers of Matewan
led the County Democrats, and they were not only considered locals but
had also been linked by marriage to the Hatfields.[40]

In addition to precincts Magnolia and Kermit (Mingo's traditional Republican strongholds), only four precincts remained loyal to
the Republicans: War Eagle, Lick Fork, Thacker, and Cinderella. The
Republican attributed the Republican hold in these precincts to the loyalty of the African-American voters. The domination of the county's other
fifteen precincts meant that for the first time in ten years, the Democrats
carried Mingo County. According to the *Williamson Daily News* there were
at least two local reasons for the Democratic victory. The first reason is
how the election was conducted. The election was quiet and well conducted because "both Democrats and Republicans agreed to post . . . guards
'over the ballot boxes.'" The presence of a bipartisan guard over the ballots meant that the Republicans could not steal the election. The second
reason is the political shift personified by the Democratic nominee for
sheriff, G. T. Blankenship.[41]

A native of Jellico, Tennessee, G. T. Blankenship had come to Matewan
in 1900 as a station agent for the N&W. He married Georgia Chambers
and joined the merchant elite of Matewan. When his brother-in-law E.
B. Chambers started the Matewan National Bank, Blankenship served as
vice-president. Along with his kinship by marriage, Blankenship's power
was built on his reputation as an honest businessman who fought both
the Hatfields' Republican machine and forged an alliance of rural district
Democrats who fought the Williamson Democrats. In the 1916 sheriff's
race, Blankenship did not stress unique campaign issues; in fact, his focus
on good roads, lower taxes, and better management of public funds almost
exactly matched the platform of Republican county court candidate S. T.
Lambert. Blankenship's main campaign promise reveals how he attracted
enough votes to defeat Wayne Damron, the Hatfield machine candidate for

G.T. Blankenship. *Matewan Democrat and Mingo County sheriff, 1916–1920.* ["Williamson Daily News," 1916, WVRHC]

sheriff: Blankenship vowed that if elected, he would "serve the masses regardless of fear or favor."[42]

To voters in Mingo in 1916, Blankenship appeared a nearly ideal candidate. The elites who were tired of the turmoil caused by the Republican and Democratic machines found in Blankenship a reputable outsider. He was not a member of a machine like his opponent and Hatfield cousin Wayne Damron, but his Chambers wife satisfied the locals for whom kinship connections were still important. For the working people, Blankenship was the first candidate for county sheriff who had actually belonged to a union.[43]

The results of the 1916 general election reflect the divergence of Mingo County and state politics. Mingo came under Democratic control. The sheriff and all three county commissioners were now Democrats. However, except for the governor's office and the House of Delegates, all of the state and federal political offices remained in Republican hands. Cornwell defeated Robinson for the governor's office by a slim margin. As Hatfield biographer Neil Shaw Penn succinctly summed up the 1916 election, "although the Republicans had lost the domination of the state government . . . the Democrats had not acquired it."[44]

The period 1912–1919 opened with the Republican Party firmly in control of Mingo County and closed with the Democratic Party equally as strong. Backed by a powerful political organization, Mingo's native son Henry D. Hatfield led southern West Virginia's Republicans to the center stage of state politics. Hatfield's forceful and autocratic leadership resolved the state's worst labor crisis and shepherded many needed reforms through the state legislature.

Hatfield's actions and methods galvanized his enemies and ensured the election of the only Democratic governor between 1896 and 1932. Ironically, his successor, John J. Cornwell, was not unlike Hatfield in several respects. Cornwell also possessed an imperious and domineering personality; those unwilling to be won over were subject to destruction, not accommodation. Cornwell juggled a similar dual identity—he was a corporate-supported man of the people. Cornwell also advocated a legislative program that highlighted Progressive reform. However, by the time labor strife reappeared during Cornwell's tenure, the changed times exposed his anti-union, pro-industry loyalty. Unlike Hatfield, who had compelled the two opposing sides to compromise, Cornwell facilitated the defeat of the miners and the near death of their union.[45]

As we have seen, the 1916 general election heralded the eclipse of the Hatfield machine and the ascendance of the Matewan Democrats. However, no single faction of either party established absolute primacy over Mingo. Williamson had remained Democratic until the last year of the Hatfields' reign. Even in defeat, between 1916 and the end of 1919, Republican factions in Williamson and Matewan still bedeviled the Democrats. Because only one state and national midterm election occurred during the Great War period, municipal and county politics dominated the non-war local news. Mingo's traditional rivalries and tensions persisted despite the election reforms of the Hatfield administration and calls for wartime unity. Chief among these were the bitter fights between the Republicans and Democrats for control of the municipal and county gov-

John J. Cornwell. *Democrat and West Virginia governor, 1917–1921. [WVRHC]*

107

ernments, the factional struggles in both parties, and the turbulent chaos of election day in Matewan.

Between 1917 and 1919, the county-level Republican and Democratic rivalry resulted from an intertwined struggle over control of both Williamson's new commission government and the county court. The Hatfield machine's strategy preserved its influence over Williamson's city government until the first municipal elections under the 1915 charter in July 1917, but the November 1916 general election broke the Republicans' hold on the county court. The county court defeat proved especially harsh because all three commissioners' seats were at stake. The Democratic court elected in 1916 consisted of Alex Bishop, a long-standing member of the City Ring, and two coal company executives, Harry G. Wilburn and L. E. Armentrout. Armentrout was superintendent of the Borderland Coal Company, one of the largest operations in Mingo County.[46]

The new court wasted little time in challenging the Republicans' waning control over the county. Within a month after the election, the new court sought to take office early based on their assertion that as elected commissioners their authority superseded the lame-duck Republican commission which by the fall of 1916 consisted of members who had been appointed to fill the seats of elected members. For the last two months of 1916, both courts held sessions and attempted to govern the county, although the old court was not legally required to disband until January 1, 1917.[47]

In addition to sweeping the county court elections, Mingo's Democrats also elected Matewan Democrat G. T. Blankenship county sheriff. These dual victories should have afforded the party virtually unassailable domination over Mingo County. However, the factional rift between Wallace J. Williamson's City Ring and the County Democrats led by Blankenship and his Chambers in-laws resurfaced even before the new court established its authority. As "the only out-and-out Williamson supporter to be elected to county office," Alex Bishop's being selected as president of the county court provided further critical access to power for the "City Ring." From December 1916 until December 1918, Sheriff Blankenship and the Bishop court squabbled frequently. At issue: who held the reins of power in Mingo, the Matewan-led County Democrats or the City Ring?[48]

The first conflict arose from a budget dispute that took nearly a month to resolve. In early December 1916, the new sheriff submitted an $8,000 budget to the county court for approval. The county court authorized only $5,800. To Blankenship's embarrassment, the court also "flatly refused" to pay for the purchase of a horse for the sheriff. The list of Blankenship's intended staff helps explain the court's behavior. Blankenship's budget was intended to cover the salaries of three deputies, one jailer, and an office manager: W. O. Porter, chief deputy; William Brewer, field deputy for the western section of the county; Lee Chambers, field deputy for the eastern portion; William "Red Bill" Damron, jailer; and Toney Webb, sheriff's office manager. Porter was the only Williamson-affiliated member of Blankenship's intended staff. Both Damron and Webb had recently switched allegiances from the Republican Party. Blankenship sought to reward their assistance to him in the campaign; the county court undermined Blankenship's ability to do so by gutting his budget.[49]

A statewide fiscal reform further complicated the standoff between Mingo's sheriff and its county court. As of January 1, 1917, West Virginia's county officers would no longer receive percentages of revenue collections as payments for their services. From January 1917, county officers received fixed salaries, and the money they collected from taxes and fines was considered "public money belonging to the county fund." The advent of salaried pay combined with the county court's budgetary restrictions altered the make-up of Blankenship's staff. Bill Damron, who already held the position of deputy county court clerk, "declined" the now lower-paying jailer's job. The position went instead to W. O. Porter, the lone City Ring man on the sheriff's staff. Thus, the first round of the conflicts between Blankenship and the Bishop court ended in tenuous compromise.[50]

The spring and summer of 1917 precipitated new confrontations in Mingo County politics. The first municipal election in Williamson since the 1915 charter reform highlighted not only the ongoing battle between the Republicans and Democrats but also the persistent fissures within each party. As mentioned earlier, the 1917 Williamson municipal election was a critical fight for Mingo's Republicans. Fallout from both the heavy-handed rule of the Hatfield machine and the revolts against it still shadowed the party.

Henry D. Hatfield's single-minded pursuit of political power had precipitated a crisis in Republican state and local politics. In addition to alienating the coal industry with his advocacy of many reform measures, he also weakened ties with some of his oldest political allies. Since the founding of Mingo County, liquor sales and distribution had been a mainstay of Republican power. R. W. "Bob" Buskirk, one of the wealthiest and most powerful native Republican elites, was Mingo County's primary purveyor of alcohol. West Virginia's prohibition of alcohol, which had taken effect in the summer of 1914, forced Buskirk, the builder of the Urias hotel and "founding father of Matewan," to abandon his base of operations. After supporting the Democratic slate of candidates in the 1916 election, Bob Buskirk abandoned the Republican Party and joined Mingo's Democratic Party in the spring of 1917. Buskirk's departure from Matewan and Republican ranks further weakened a party already devastated by the Hatfields' strategic blunders and ongoing legal troubles.[51]

Despite the obvious involvement of Hatfield machine hacks in the election fraud at the War Eagle and Rockhouse precincts in 1914, the defendants, who included Mingo County Republican Party chairman Greenway W. Hatfield, continued to elude legal punishment. The distraction of recurrent trials encouraged rumors about the Hatfields' slipping power in county politics. Within a week of Buskirk's defection from the party, the *Mingo Republican* reported that stories had been circulating about Greenway Hatfield's possible replacement as Republican county chairman by Harry Scherr, a well-connected Williamson coal attorney. The newspaper countered the rumor with a charge that it had been started by another Republican defector Toney Webb, whose kinsman was a candidate in the upcoming Williamson municipal election.[52]

The Williamson municipal election fired political controversy months before it actually occurred in the summer of 1917. Borrowing a page from the tactics used by his Hatfield kinsmen and frequent adversaries, Judge James Damron orchestrated the reorganization of the primary ballot entries in an effort to boost the reelection chances of his brother-in-law and mayoral hopeful J. M. Studebaker. Had the commission candidates' names appeared on the ballot as they had in previous elections, the names of Studebaker and W. O. Porter, his chief Democratic rival, would have been arranged so that a vote for one would have canceled a vote for the other. Damron's machinations on behalf of Studebaker resulted in Porter's name being arrayed against that of Gail T. Dudgeon, the Hatfield machine candidate. To a degree, Damron's strategy worked, as Studebaker garnered more votes than Dudgeon, but with only 325 votes, he came in a distant second to Democrat W. O. Porter, who received 478 votes.[53]

The runoff primary proved Studebaker's weakness and forced Damron to appeal to party boss Greenway Hatfield for an end to the factional fighting. The deal offered to Hatfield by Damron and Studebaker reveals the importance of political office and patronage to local party organizations. Damron and Studebaker proposed replacing Williamson's chief of police, deputy county clerk, and current street commissioner with "friend[s] of Greenway Hatfield's." Buoyed by the second mistrial verdict in the Rockhouse Case less than a month before, Hatfield flatly rejected Damron's offering. Hatfield explained by reminding the *Mingo Republican*'s readers that Studebaker had conspired to "throw [a Republican mayor] out of office" and then had voted to appoint Democrats to city assessor and city attorney. Hatfield noted that Studebaker had also not objected to the firing of Williamson's only African-American employees. Hatfield concluded by adding that the appointments specifically offered by Damron were "all to the advantage of the party traitors," and that a Studebaker victory would actually be a Democratic victory. With the Republicans unable to compromise, Democrat W. O. Porter won the mayoral seat in the June 1917 municipal election. The 1919 municipal election was also hotly contested, but again, Porter emerged the victor.[54]

Local politics throughout Mingo County remained fractious in the entire period preceding the Matewan Massacre. The faction wars between the City Ring and the County Democrats counterbalanced the disintegration of the Hatfield machine's stranglehold on the county's Republican Party. As a result, groups within each party constantly shifted allegiances and formed new alliances in their efforts to seize control of county politics. The annual Matewan municipal elections and the 1918 midterm election highlight the personalities and issues that influenced the temper of Mingo County politics when the United Mine Workers of America launched their organization drive in 1920.

The home base of Mingo County's most powerful Republican faction *and* the insurgent Democratic faction, Matewan proved to be the county's most volatile political arena. Election day in Matewan perennially featured drunkenness, fistfights, and charges of illegal voting. While state and national attention had been drawn to events in Matewan on general election days, the true nature of Mingo County politics was most clearly reflected in Matewan's municipal elections. Held in January, months before the primaries, Williamson's municipal elections, and the November general elections, Matewan's municipal elections were the harbingers of the political year to come.

The 1917 municipal election in Matewan proved to be the sole electoral victory in a year of disappointing political setbacks for the Hatfield machine. During the first week of January, the brother of Henry D. and Greenway, A. B. (Andy Barrett) Hatfield, won reelection as Matewan's mayor, along with the rest of his slate, the Independent League. Hatfield's opponent in the mayoral race was Cabell Testerman. Hatfield had won three successive municipal elections and served as mayor from 1915 to 1917. The Hatfield machine's defeats at the county and state level in November 1916 and Buskirk's defection in March 1917 undermined A. B. Hatfield's ability to retain power in Matewan; an era was about to end.[55]

The Matewan 1918 municipal election illustrated how far the Hatfield influence had declined. The Hatfield machine traditionally benefited from election day violence and chaos, but the election of Matewan Democrat G. T. Blankenship as sheriff deprived the Republicans of that advantage. Although activity around the election booths started out "lively," the arrival of Blankenship and his deputies calmed the situation, and the election proceeded smoothly. According to the *Williamson Daily News,* the election was "between the highest citizenship of the town on one side and a bunch of fellows on the other who are continually causing trouble." The 1918 victory of Cabell Testerman and of the Citizens' Party over A. B. Hatfield and the Independent League inaugurated three years of Democratic rule in Matewan.[56]

The accumulation of political defeats and the ongoing election fraud cases inspired the *Williamson Daily News* to observe, "We had thought the Republican Party in Mingo County was dead, eternally dead for all time to come." What the *Daily News* did not state explicitly was that local Democrats believed that the local Hatfield machine had been crushed.

However, in early 1918, when the Republican Executive Committee over-hauled its membership, the Hatfield machine remained an entrenched force in the county's Republican leadership. Seven of the committeemen were economically and/or socially linked to the Hatfield machine. The resurgence of the Hatfield machine led the *Williamson Daily News* to declare that Mingo's Republican Party was "still kicking" and that the new Republican committee was "a most harmonious gang."[57]

The power of the Hatfield machine had reached a low point when it lost control of Matewan, its traditional powerbase, in January 1918. The retention of its majority in the county organization in March marked the machine's resurgence. Shortly before the May primary, the third and final trial of the Rockhouse Case ended in acquittal. The resolution of these cases liberated Greenway Hatfield and several of his associates from the threat of prison. The 1918 midterm election offered only six contests: a United States senatorship, a United States congressional seat, seats in the state Senate and House of Delegates, and two county positions (short term commissioner of the county court and county superintendent of schools), There was little for the Hatfield machine to gain. Overshadowed first by the war, and then by the influenza epidemic, the 1918 campaign in Mingo County was muted and lacked the fire of previous contests.[58]

The dominant reason for the low-key campaign was preoccupation with the war effort. Raucous campaigning would have been viewed as unpatri-otic. The *Williamson Daily News* published articles on national politics in which President Wilson urged unity among party leaders because discord only encouraged the Germans. The campaign advertising that appeared in Mingo County reflected the overwhelming importance placed on the war over local issues. West Virginia's Democratic candidate for U.S. Senator, Clarence W. Watson, conducted his campaign while still in the Army and his advertisement included a photograph of Watson in uniform. The campaign literature of Wells Goodykoontz, the Republican candidate for U.S. Congress from West Virginia's fifth district, focused on how, after the war, he would work to right the wrongs perpetrated by southern Democrats during the war. Goodykoontz believed the Democrats had unjustly scape-goated the coal industry and over-regulated it during the war, while allow-ing Southern agricultural interests to benefit from wartime inflation.[59]

However, neither Watson's nor Goodykoontz's advertisements appeared until just a few weeks before the election. In fact, the entire campaign had been compressed. In Mingo County the primary was delayed until August, and the candidates waited until the fall to actively campaign. The outbreak of influenza then undermined the 1918 September-October campaign. The state health department banned all public meetings, and candidates for office, such as R. L. Harris, Democratic candidate for the county court seat, restricted their canvassing for fear of spreading contagion. Some candidates, like Evan A. Justice, Mingo's Democratic candidate for state senate, actually fell ill with "the Flu" and were unable to campaign.[60]

Held on November 5, the 1918 election bitterly disappointed Mingo's Republicans. Only two Republicans, Wells Goodykoontz and Dr. William York, defeated their Democratic opponents. Goodykoontz lost in his home county of Mingo but carried enough votes in the district's other counties to win. York vanquished influenza-stricken Evan Justice for Mingo's state Senate seat. The victory of Democrats Harris, Evans, and Thomas exposed the continued weakness of the mainstays of Mingo's Republican Party, the coal industry Republicans, the Hatfield Republicans, and the Republicans of Matewan. C. M. Gates, the Republican candidate for county commissioner and general superintendent of the Crystal Block Coal & Coke properties, lost to Mingo native R. L. Harris. Harris's political advertisement underscored the class aspect of the nativity issue. He declared: "I . . . have always lived with you and labored with our people . . . for the good of all alike." This assertion implied that Gates, who had been superintendent of Crystal Block for "several years" but had been a county resident for just a few months, could not and should not represent the people of Mingo County on the county court. C. M. Whitt, the Hatfield machine candidate for house of delegates, lost to the young and virtually unknown Rice Thomas. In the race for county superintendent of schools, Matewan Republican and Hatfield kinsman Wilson Chafin even failed to defeat Floyd Evans, a Democrat from Kermit, a remote town the Republicans had incorporated and named for the son of President Roosevelt less than a decade before.[61]

Still, the Democratic victory over the Republicans in the county elections did not swing the balance of power to the City Ring. R. L. Harris of

Chattaroy and Floyd Evans were Democrats from the County faction and, as such, natural allies of Blankenship. Harris' anti-coal operator advertisement also indicated that it was unlikely he would follow the lead of the Williamson Democrats, most of whom were coal capitalists themselves. Rice Thomas, the winner of Mingo's House of Delegates race, was the sole Democrat who hailed from Williamson.[62]

The last significant political event of 1918 illuminates the consolidation of the County Democrats' influence over Mingo's politics. Sometime between the end of November and early December, county court commissioner Harry G. Wilburn resigned, and E. B. Chambers, Sheriff G. T. Blankenship's brother-in-law, succeeded him. The *Williamson Daily News* remarked that Chambers had repeatedly refused various political honors and, now that he had accepted a position, would represent "all the people in Mingo County, regardless of whether they are Democrats or Republicans." Nearly two years into his term as sheriff, Blankenship had an ally on the county court.[63]

Within a month of Chambers' ascendance to the county court, Matewan held its 1919 municipal election. Although Chambers' and Blankenship's fellow Democrat and ally Cabell Testerman won his second consecutive election, the election day commotion foreshadowed conflicts to come. Moreover, behind the scenes, the Williamson Democrats regrouped, and by the end of 1919, all of Mingo's political factions were gearing up for the campaigns of the 1920 general election. Despite the plea from the *Williamson Daily News* that local politicians "give the people of Matewan a square election and stop endeavoring to gain power by manipulation," the 1919 Matewan municipal election served as the opening salvo of several years' discord.[64]

Former mayor and current Matewan chief of police A. B. Hatfield had marshaled a force of men to help him regain the mayoral seat. Described as a gang of "desperado bootleggers" and "mountaineer ruffians from the wilds of Blackberry Creek," Hatfield's men tried to "bull-doze" the election for their boss. The situation reached critical mass when Chief Hatfield ordered an African American from Red Jacket to vote over the man's own objection that he was not a legal voter. Hatfield reiterated his order and added, "nobody dares cross my path—I am a Hatfield." Sheriff

Blankenship, who had brought his deputies to Matewan to watch over the election, ordered the man arrested. Hatfield demanded that the man, later identified as Will Dudley, a recent parolee from a Virginia penitentiary, be released. Blankenship refused to bow to Hatfield's demand. Chief Hatfield and "his blood-thirsty bunch" started, but stopped when a constable put a "hand-cannon" to Hatfield's "thinking cap." Hatfield's men ran, Hatfield backed off, and Blankenship restored order.[65]

The armed standoff in Matewan did not end the town's political strife. After Hatfield retreated on election day, Blankenship rounded up suspected illegal voters. Will Dudley and six other African-American men ultimately appeared in magistrate's court. Dudley pleaded guilty, and Republican county chairman and Chief Hatfield's older brother Greenway Hatfield "gave surety" for the other men. The *Williamson Daily News* condemned the Hatfields' behavior as "the final chapter" in Matewan's "corrupt elections" but lauded Sheriff Blankenship's actions: "we are glad to see that the day is dawning when we have an official who says he will protect us with all his power . . . the citizens are back of you."[66]

The Hatfields' political trouble and public embarrassment continued in the spring of 1919. Ex-mayor and chief of police A. B. Hatfield and N. L. Chancey, who served as town recorder during Hatfield's tenure, faced more public censure. According to a story published in the *Daily News*, Hatfield and Chancey had "tampered" with Matewan's municipal records and "fixed things to suit their taste." The article specifically charged the pair with excising pages from the town record "when sentiments became so hot that an investigation was planned." However, Chancey had already fallen into disgrace for "filching" funds from the Magnolia district board of education, while Hatfield enjoyed his brothers' protection. Despite the accusations regarding his mismanagement of Matewan's resources, A. B. Hatfield did not fade from the political scene. By 1920, he was serving as a magistrate for the Magnolia District.[67]

The grudges that accumulated in the annual fight between Hatfield and his allies and Blankenship, Testerman, and the Chambers' for control of Matewan directly affected the escalation to violence on May 19, 1920. Matewan and Mingo County politics from 1912 through 1919 had remained just as volatile as they had been early in the county's history.

The one constant theme in all of the maneuvers of Mingo's leaders was the ability of the machines led by the native elites to survive the efforts of reformers and insurgents. The escalating pressures on the machines in this period, however, precipitated profound shifts that directly affected the events of 1920. In particular, the machinations of 1915–1916 first cost the City Ring Democrats control of the county seat and then cost the Hatfield Republicans control of the state and Mingo County. The end result was the elevation of County Democrat G. T. Blankenship to sheriff. A former railroader and union man, Blankenship would be in office when union organizers arrived in 1920. When he denied Albert Felts legal assistance for the Stone Mountain evictions on May 19, 1920, Felts turned to Blankenship's old nemesis, A. B. Hatfield, for aid. As time and events would show, this local political turmoil only exacerbated Mingo's uncertain economic future.

NOTES

1   Karr, 63; Source for "outs": John Alexander Williams, *West Virginia and the Captains of Industry* (Morgantown: West Virginia University Libraries, 1976), 97; *Williamson Daily News*, October 28, 1914; Frederick Barkey, "The Socialist Party in West Virginia from 1898 to 1920: A Study in Working Class Radicalism" (Ph.D. diss., University of Pittsburgh, 1979), 175; Source for failure of primary and ballot reform: Richard Hofstadter, *The Age of Reform: From Bryan to FDR* (New York: A. Knopf, 1956), 265. A more recent study of the Progressive era affirms Hofstadter's assertion. See Diner, 231.

2   *Mingo Republican*, November 1, 1912; For references to issues Hatfield raised while campaigning in other counties, see Neil Shaw Penn, "Henry D. Hatfield and Reform Politics: A Study of West Virginia Politics from 1908 to 1917" (Ph.D. diss., Emory University, 1977), 286; *Mingo Republican*, November 8, 1912.

3   See the studies of Karr, Penn, and Burckel for details.

4   A. D. Lavinder interview with Bill Taft and Lois McLean, June 22, 1973, in Matewan, Mingo County, WV, Oral History Collection, WVRHC; Frederick Barkey, 169. During the same campaign, Hatfield's cousin Don Chafin also campaigned for Logan County sheriff with a promise to get rid of mine guards. After the election, Chafin also "kept his promise" by replacing the guards with an "army of deputies." In Mingo County, Greenway Hatfield, Henry D.'s brother,

followed Chafin's example; *Williamson Daily News*, April 22, 1916.

5   Barkey, 175. This assertion is based on the candidacy of "machine" politicians in subsequent elections.

6   Source for "lack": Karr, 157 and 168–170; Source for use of violence: Karr, 144, 153–154 and Penn, 23, *Bluefield Daily Telegraph*, September 5, 1908; Source for attack on operator: Lee, 45–6 and Corbin, *Life, Work, and Rebellion*, 88; Source for attack on commissioner: Karr, 153–154 and *Williamson Daily News*, May 12, 1914.

7   *Williamson Daily News*, April 22, 1916; *Mingo Republican*, September 28, 1916. See Table 8 in this chapter for details; Venchie Morrell interview with Rebecca J. Bailey, Summer 1990 Matewan Oral History Project; Hatfield correspondence letter no.6.

8   *Williamson Daily News*, October 31, 1918; *Williamson Daily News*, November 2, 1915; "Williamson Charter," Senate Bill #199 *Acts of the Legislature of the State of West Virginia*, 1915: 476–540.

9   Bradley Robert Rice, *Progressive Cities: The Commission Government Movement in America, 1901–1920* (Austin, TX: University of Texas, 1977), 10; Source for Williamson biographical information: Ely, 150, 285, 324; Source for Stokes' biographical materials: Stokes Papers, WVRHC.

10   "An Outrage Upon Our Intelligence," *Williamson Daily News*, January 30, 1915; "Williamson Charter Bill," 477; O. H. Booten, who was selected to head the commission and serve as "mayor," also had been the editor of the *Mingo Republican; Williamson Daily News*, October 31, 1914.

11   Several Republicans in the West Virginia legislature broke ranks and joined the Democrats in opposing the charters. *Williamson Daily News*, February 4, 1915 and February 5, 1915; *Williamson Daily News*, January 30, 1915.

12   *Mingo Republican*, January 6, 1916.

13   The right of municipal home rule versus the primacy of the state legislature was a legal and political dilemma as old as the American republic. Hendrik Hartog, *Public Property and Private Power: The Corporation of the City of New York in American Law, 1730–1870* (Chapel Hill: University of North Carolina Press, 1983); The cases were: "Booten v. Pinson," "Dudgeon v. Hall," "Nunemaker, et al. v. Booten, et al.," and "Pinson v. Booten et al.," *Reports of the West Virginia Supreme Court of Appeals*, 77 (October 26, 1915–March 21, 1916): 412–442; *Mingo Republican*, January 6, 1916.

14 "West Virginia's Bribery Scandal," *Literary Digest* (April 1, 1913): 441–442, quoted in Corbin, *Life, Work, and Rebellion*, 13, note 57. Corbin does not cite the volume number.

15 Penn, 309–313.

16 Tams interview, 24, WVRHC; Penn, 247, 320; *Williamson Daily News,* June 25, 1914.

17 William D. Ord to John W. Mason, June 7, 1915, quoted in Penn, 423; Karr, 50.

18 *Mingo Republican*, October 26, 1916; Justus Collins to J. Holloway May 3, 1916, quoted in Penn, 453–454; *Williamson Daily News,* January 22, 1915; Source for the existence of a Mann-Elkins coalition: *Mingo Republican,* July 29, 1915; *Mingo Republican*, March 9, 1916, and June 1, 1916; Source for Lilly's given name: "Armistead Abraham Lilly," *West Virginia Heritage Encyclopedia*, vol. 13:2795–2796. In contemporary documents and newspapers Lilly was always referred to as A. A. or "Abe" Lilly.

19 *Mingo Republican*, December 30, 1915.

20 *Williamson Daily News*, May 27, 1916.

21 *Williamson Daily News*, May 27, 1916.

22 *Mingo Republican*, June 8, 1916. Lilly carried Mercer, Wyoming, Summers and Raleigh counties.

23 Justus Collins to I. T. Mann and Mann to Collins, June 9, 1916 and June 12, 1916, Collins Papers, WVRHC.

24 *Mingo Republican*, July 27, 1916 and *Williamson Daily News,* August 3, 1916; Penn, 463; S. D. Stokes to John W. Davis, July 7, 1916, Stokes Papers, WVRHC.

25 Henry D. Hatfield later served as U.S. senator, and Greenway served again as sheriff of Mingo. *West Virginia Bluebook,* 1928–1936.

26 Source for differences between parties: Williams, *The Captains of Industry*; The phrase "party-army" is borrowed from McKinney's *Southern Mountain Republicans*; Rice and Brown, 214; William P. Turner, "From Bourbon to Liberal: The Life and Times of John T. McGraw, 1856–1920," (Ph.D. diss., West Virginia University, 1960), 249.

27 One reason for the lack of information on the Democratic Party in Mingo County is the virtual absence of newspapers prior to 1911 (there are only scattered issues from June 1908) and the lack of manuscript collection materials from any

Mingo County Democrats for the early period from 1895–1912; *Mingo Republican*, November 1, 1912.

28 *Mingo Republican,* May 21, 1914; It is likely that W. N. Cole, the ousted Democratic chairman, was the same Cole who had started the Mattie May mine in 1905. If so, the Cole-Williamson power struggles take on a new dimension, as "Martha May" was the daughter of former Democratic sheriff, Ali Hatfield. Williamson and Matewan had been rivals since 1895 when Williamson narrowly defeated Matewan in the special election to select Mingo's county seat.

29 *Mingo Republican,* March 17, 1911; *Mingo Republican,* May 21, 1914; *Mingo Republican,* October 15, 1914, and October 22, 1914.

30 *Williamson Daily News,* November 12, 1914; *McDowell Recorder,* May 24, 1918; *Mingo Republican,* September 14, 1916. Reprint of editorial from the *Huntington Herald Dispatch.* Two separate election fraud cases from Mingo reached Federal District court between 1916 and 1918.

31 *Williamson Daily News,* June 2, 1916. On May 18, 1920, the day before the Massacre, the State Supreme Court denied an effort to clarify Mingo's voter registration rolls before the primary.

32 *Williamson Daily News,* June 2, 1916; *Williamson Daily News,* 5 June 1916.

33 *Williamson Daily News,* June 5, 1916.

34 *Williamson Daily News,* November 2, 1916; *Mingo Republican,* April 12, 1917.

35 *Williamson Daily News,* June 5, 1916; *Mingo Republican,* September 14, 1916. In this same issue, an item written by J. Jerome Haddox and first printed in the *Lincoln Republican* describes a 1914 election abuse committed by Democrats at Sharples precinct in Logan County; *Mingo Republican,* June 15, 1916.

36 *Mingo Republican,* September 28, 1916 (reprint from *Huntington Herald Dispatch*); In the *Williamson Daily News,* June 6, 1916, it was reported that only 40 out of 157 registered voters actually cast ballots. "The rest . . . [were] copied into the poll book in alphabetical order as they appeared upon the registration books"; The story of mule and bulldog voting in Mingo County is also corroborated by oral testimony, and in the case of the bulldog, a contemporary newspaper: *Williamson Daily News,* August 8, 1914.

37 *Mingo Republican,* October 5, 1916.

38 Source for "Rockhouse Case": *McDowell Recorder,* May 24, 1918; Source for "menace" and "combination": *Mingo Republican,* October 12, 1916; The

"Rockhouse Case" ended in acquittal on May 21, 1918.

39 *McDowell Recorder,* May 24, 1918; *Mingo Republican,* November 2, 1916.

40 *Williamson Daily News,* November 8, 1916. The banner headline read, "MINGO COUNTY REDEEMED AT LAST!"; *Mingo Republican,* March 1, 1917; *Mingo Republican,* October 12, 1916.

41 *Williamson Daily News,* November 8, 1916; *Mingo Republican,* March 1, 1917; *Mingo Republican,* October 12, 1916.

42 *Williamson Daily News,* "G. T. Blankenship political advertisement," April 8, 1916; Information about Blankenship taken from Matewan National Bank letterhead on correspondence dated March 20, 1916, Stokes Papers, WVRHC; Source for Lambert: *Mingo Republican,* June 5, 1916.

43 Source for Blankenship's "connections": Hatfield correspondence, letter no. 27; Source for attitudes towards politics: Rose Nenni Ore interview with John Hennen, Summer 1989 Matewan Oral History Project; Source for assertion about common Progressive Era phenomenon: Charles N. Glaab and A. Theodore Brown, *A History of Urban America.* New York: Macmillan, 1983 [1967], 208; Source for "working": *Williamson Daily News,* April 8, 1916.

44 *Williamson Daily News,* November 8, 1916; Lucy Lee Fisher, "John J. Cornwell, Governor of West Virginia, 1917–1921," *West Virginia History* 24 (April 1963/July 1963): 274; Penn, 472. Republican James Hughes held on to his fifth U.S. Congressional district seat. Republican Howard Sutherland defeated W. E. Chilton for United States senator. Four out of the six Congressional seats were filled by Republicans. Republicans dominated the state senate by ten and completely controlled the board of public works.

45 Fisher, 263, 269, 272–273, 276.

46 *Mingo Republican,* December 7, 1916.

47 *Mingo Republican,* December 28, 1916.

48 The scarcity of pre-1911 primary sources complicates efforts to trace the internal divisions within Mingo County's Democratic Party. It is likely that the rivalry dated as far back as the fight for Mingo's county seat in 1895; Source for "out": *Mingo Republican,* January 11, 1917; Source for "squabbled": *Mingo Republican,* January 16, 1916; *Mingo Republican,* December 7, 1916, October 25, 1917.

49 *Mingo Republican,* December 14, 1916, and December 21, 1916. Toney Webb had resigned his position with the Red Jacket Consolidated Coal & Coke Company in anticipation of his "reward."

50 *Mingo Republican,* December 28, 1916; *Mingo Republican,* January 4, 1917.

51 "Big Dry Victory," *Mingo Republican,* November 8, 1912. Inspiration for looking for a connection between mining and the prohibition amendment came from Michael Nash, *Conflict and Accommodation: Coal Miners, Steel Workers, and Socialism, 1890–1920;* Hatfield correspondence, letter no. 17; Johnny Fullen interview; *Mingo Republican,* March 8, 1917.

52 *McDowell Recorder,* May 21, 1918; *Mingo Republican,* March 1, 1917.

53 *Mingo Republican,* March 29, 1917 and April 5, 1917; *Mingo Republican,* May 31, 1917; *Mingo Republican,* April 26, 1917.

54 *Mingo Republican,* May 31, 1917. "Red Bill" Damron was James Damron's brother. *Mingo Republican,* June 1, 1916; *Mingo Republican,* May 3, 1917. The second mistrial in the "Election Fraud Cases" was declared on April 29, 1917; *Mingo Republican,* September 20, 1917; *Williamson Daily News,* June 6, 1919.

55 *Mingo Republican,* January 11, 1917. For the municipal elections, the Republicans called themselves the "Independent League," and the Democrats were the "Citizens' Party"; Identified by the *Williamson Daily News* as the brother of Greenway and Henry D. Hatfield, A. B. (Andy Barrett) does not appear in either contemporary or current genealogies of the Hatfield family. The logical explanation for his omission might be illegitimacy, which was not unknown in the Tug Valley or among the Hatfields; *Williamson Daily News,* June 5, 1916.

56 *Williamson Daily News,* January 12, 1918; *Williamson Daily News,* January 5, 1918. Testerman won the 1918, 1919 and 1920 Matewan municipal elections.

57 *Williamson Daily News,* February 23, 1918.

58 *McDowell Recorder,* May 21, 1918.

59 *Williamson Daily News,* February 2, 1918; *Williamson Daily News,* October 31, 1918; *Williamson Daily News,* November 5, 1918.

60 *Williamson Daily News,* July 18, 1918; State of West Virginia Department of Health, *Biennial Report of the West Virginia State Department of Health, 1919–1920* (Charleston: Tribune Printing Company, 1920),12; Source for "labored: *Williamson Daily News,* October 31, 1918.

61 *Williamson Daily News,* January 9, 1918; Source for who won which counties: *Legislative Handbook and Manual of the West Virginia Legislature 1919,* 336; Justice was a Blankenship ally; *Mingo Republican,* February 15, 1917; *Williamson Daily News,* October 31, 1918; *Williamson Daily News,* November 9, 1918.

62   *Williamson Daily News*, October 31, 1918.

63   S. D. Stokes to G. R. C. Wiles, March 31, 1919, Stokes Papers, WVRHC; *Williamson Daily News*, December 12, 1918.

64   *Williamson Daily News*, December 12, 1918.

65   *Williamson Daily News*, January 4, 1919. The newspaper does not identify the constable who stopped Hatfield. Both constables in Magnolia District (Harry Chafin and A. L. "Al" Hoskins) were Republicans.

66   *Williamson Daily News*, January 7, 1919; *Williamson Daily News*, January 4, 1919.

67   *Williamson Daily News*, April 22, 1919. One of the other targets of the accusatory article later asserted that its author was Toney Webb, one of Matewan's more infamous Republicans turned Democrat; *Williamson Daily News*, April 24, 1919. The charge concerning Webb requires then that the accusations concerning Hatfield and Chancey be viewed cautiously. Webb could have been revealing information he had been privy to prior to his defection or fabricating a believable story to undermine his former compatriots. The possibility of purposeful record destruction reminds one of the complexity of politics in Mingo County and southern West Virginia; it also enhances the pathos of the loss of historical records to natural disasters like the floods of 1977 and 1984; This assertion is based on A. B. Hatfield's ability to retain positions of influence despite electoral defeat. Between 1918–1920, Hatfield went from Matewan's mayor to chief of police to Magnolia District magistrate.

3

THE WILLIAMSON-THACKER
COALFIELD FALLS BEHIND

"See, we didn't have any big coal operators."
– *Howard Radford*[1]

TWENTY YEARS AFTER ITS OPENING, the Thacker coalfield, now referred to
as the Williamson-Thacker coalfield, appeared to be just another of southern
West Virginia's expanding coalfields. However, superficial commonalities
between it and its larger neighbors to the east and newer neighbors to
the north and south masked the ways in which the Williamson-Thacker
coalfield had fallen and would fall even further behind the rest of the
region's anti-union coalfields. By the end of World War I and the beginning
of the 1920s, what set the Williamson-Thacker coalfield apart made it the
ideal battleground for control of the Appalachian coal industry.

Between 1912 and 1916, coal production in the Williamson-Thacker
coalfield increased. However, expansion did not translate into greater economic security for either the companies or their employees. Despite the
increased presence of more highly capitalized companies and the implementation of several "reforms," discontent in the county's mining population rose. The outbreak and resolution of West Virginia's first mine war,
while it did not directly affect the Williamson-Thacker coalfield, altered
the course of labor relations in southern West Virginia and in turn assured another collision between the United Mine Workers of America and
the anti-union operators of southern West Virginia.

In the years before the United States entered World War I, the boom in
coal production in Mingo County had slowed, and the number of skilled
miners had declined dramatically. Continued expansion in the number
of mines in the county did not necessarily mean that the miners them-

selves benefited from better work opportunities. Yearly fluctuation in the number of days the mines operated indicated that although the miners received higher daily wages, steady employment remained elusive. Moreover, because companies offset wage hikes by boosting prices in the company stores, housing rents, or the services provided by the company doctor, the miners' standard of living did not improve.[2]

The consolidation of smaller and/or older mining operations by large corporations received a great deal of press coverage in the 1912–1916 period. During congressional hearings into the consolidation of economic power on Wall Street, federal advocate Samuel Untermyer revealed that the Red Jacket Consolidated Coal & Coke Company, one of the largest operations in Mingo, was among the "coterie of banks and coal companies" controlled by Isaac T. Mann of Bramwell, Mercer County. According to Untermyer, a Mr. Stotesbury, a member of J. P. Morgan & Company, was also a member of Red Jacket's board of directors and director of two companies that held a significant number of Red Jacket's bonds. Under two different corporate monikers since 1902, Red Jacket had, in addition to opening new mines, absorbed several of the county's oldest mines, including the Lick Fork and Grapevine mines. Morgan & Company controlled at least one other mining operation in Mingo County by 1912—the White Star Mining Company, which, along with several other sister mines, provided the coal for Morgan's White Star shipping line.[3]

In addition to the Morgan interests, also known as U.S. Steel, other large corporations pursued developments in the Mingo section of the Tug River. On September 16, 1913, Thomas H. Claggett, an investigator for the Pocahontas Coal & Coke Company, reported that the Berwind-White interests were buying up more acreage on Pond Creek and Tug River. Only the mouth of Pond Creek lay in West Virginia; the Berwind-White lands were in Kentucky, in the new Pond Creek coalfield. The opening of this Kentucky coalfield profoundly affected Mingo's future. By 1917, when the Pond Creek coalfield opened, the Williamson-Thacker coalfield was surrounded by coalfields that were the focal points of intense and heavy development by the most powerful corporations in the Appalachian coal industry. To the southeast, U.S. Steel controlled most of the Pocahontas coalfield. To the north-northeast the Island Creek Coal Corporation dominated the Guyan (Logan) coalfield, and after 1912, the Pond Creek coalfield would be divided among the Berwind-White, Island Creek, and Consolidation Coal interests. As in the case of the opening of the Guyan (Logan) coalfield in 1904–1905, the rapid development of the Pond Creek coalfield underscored the position of the Williamson-Thacker coalfield as a marginalized, peripheral field. The big corporations developed holdings in Mingo but never on the scale that occurred in surrounding counties. More important, the Williamson-Thacker coalfield also never became the primary base of operations for any of them. By the end of World War I, Mingo County and the Williamson-Thacker coalfield would fall into the shadow of its neighbors, and it would take violent conflict to lift the pall and focus national attention on the relative backwater.[4]

A decade before the Williamson-Thacker Strike of 1920–1922, a contract dispute in Kanawha County dissolved into one of West Virginia's worst episodes of labor strife. Also known as the first West Virginia Mine War, the Paint Creek and Cabin Creek Strike of 1912–1913 anticipated the Williamson-Thacker Strike in several ways. Dissatisfied with the progress of contract negotiations, the union miners of the Paint Creek district walked out of the mines in April 1912. Quickly joined by several thousand miners from neighboring Cabin Creek, the strike provided the United Mine Workers of America with yet another opportunity to launch a unionization drive in southern West Virginia. However, the conflict also set in motion a chain of bitter and often violent confrontations that would not end for decades.[5]

Although the operators of Paint Creek had acquiesced to unionization in 1902, by 1912 they were determined to drive the United Mine Workers of America from the Kanawha coalfield. While the miners' union raised funds for the strike and sent high-ranking officials to the field to assist the local officials of District 17, the operators also amassed a war chest and braced for fierce resistance. The operators fortified their properties, imported several hundred Baldwin-Felts detectives, and armed them with machine guns. By June, the operators had gone on the offensive, unleashing the mine guards on the strikers.[6]

Throughout the rest of 1912, the campaign of violence continued. In one instance, a specially equipped train peppered a strikers' tent colony with machine gun fire, killing at least one inhabitant. At a loss to deal with the operators' virulent opposition, the national union leadership's enthusiasm for the strike fell as the cost of the strike mounted. At least two attempts to broker a compromise were flatly rejected. Frustrated by the national officials' waning support and incensed by the vacillating ineffectiveness of their leaders in District 17, the striking miners turned to the radical wing of the local Socialist party.[7]

Galvanized by the financial and moral support of their new allies, the rank-and-file miners of Paint and Cabin Creeks struck back at the operators. Armed with "bootlegged" guns, striker-snipers shot at mine guards and operators, while others blew up tipples and attacked coal trains moving through the district. At the same time, the state's Socialist press publicized the atrocities visited upon the miners. The operators pressured West Virginia's governor, William E. Glasscock, to act. Derisively referred to as "crystal-peter" by union activist Mary Harris "Mother" Jones, Glasscock declared martial law and authorized violators to be arrested and tried by military tribunal. After several hundred miners were taken into custody and many were sentenced to penitentiary terms, confrontations between the strikers and the mine guards finally subsided during the winter of 1912–1913.[8]

When only miners were subjected to this harsh justice, which baldly violated United States Constitutional law, union attorneys fought back. The West Virginia State Supreme Court of Appeals endorsed the governor's proclamation and upheld the martial law convictions, but the tide of the strike turned yet again. After martial law focused national attention on West Virginia, the United Mine Workers of America, the new governor (Henry D. Hatfield), and the operators once again sought to resolve the conflict. Hatfield met with union officials and hammered out a compromise which called for a nine-hour work day, the miner's right to select a checkweighman, semi-monthly pay, and no discrimination against union members. When he presented the new terms to an operators' delegation, Hatfield won their agreement by physically assaulting the most recalcitrant operator. The Hatfield Agreement ignored the striking miners' two most important demands, complete recognition of the union and the abolition of the mine guard system.[9]

To avoid open confrontation with the local miners, Hatfield encouraged the union to break with the tradition of submitting the proposed contract to a referendum vote. Hatfield and the union leaders instead called a convention of selected miners' delegates to vote on the settlement. Despite the arrest of delegates who opposed the contract, the remaining miners' representatives debated for three days before approving the contract. Hatfield's machinations backfired, however, when the rank-and-file miners of Paint Creek and Cabin Creek learned that the Hatfield Agreement still did not address the issues of union recognition and the abuses of the mine guard system.[10]

Having gained the compliance of the union leadership and the local operators, Hatfield had no patience for the miners' obstinacy. He issued a 36-hour ultimatum for the resumption of work at the mines and sent soldiers to "escort" the miners to work. When two Socialist newspapers criticized Hatfield's action in the settlement proceedings, he ordered a raid on their offices during which their presses were destroyed. Once again national attention focused on West Virginia, and the United States Senate launched an investigation of the strike and the tactics utilized to end it.[11]

After several months the Senate hearings ended with a denunciation of the actions of Governor Hatfield, the military authorities of West Virginia, and the operators of the Kanawha field. The Senate committee also asserted that in their attempts to force the miners to end the strike, Hatfield and the operators had committed "numerous violations of the U.S. Constitution and several West Virginia state laws." The rebellious miners capitalized on the uproar caused by the Senate hearings to press for acknowledgment of their original demands. Under pressure, Hatfield and the operators capitulated. More than a year after the conflict had begun, the miners of Paint Creek and Cabin Creek won an agreement that guaranteed recognition of their union and an end to their oppression by the mine guard system.[12]

Although the miners of the Kanawha coalfield preserved and even expanded their rights as union members, the conflict itself precipitated a series of actions by both the union and the anti-union operators of southern West Virginia that bore fruit when the miners and operators of the Williamson-Thacker coalfield squared off in 1920. After the victory of the Kanawha miners in 1913, the anti-union operators of southern West Virginia launched an organization initiative of their own to protect their interests against future assaults from the union. The operators prepared to resist further encroachment by the union and hoped for an opportunity to "roll it back" from West Virginia's coalfields. More secure in their foothold in Kanawha County, the United Mine Workers of America continued to look for opportunities to finally organize all of southern West Virginia.[13]

After their defeat in the Paint Creek-Cabin Creek strike, the operators moved to fortify their position against union attack. First, after recognizing "that in union there is strength," the operators increased the number of associations organized by coalfields and created a statewide operators' association. This movement was led by Z. T. Vinson, whose confrontations with UMWA attorney Harold W. Houston during the Paint Creek hearings only confirmed his long-held belief that all challenges to industrial interests in West Virginia were backed by the forces of socialism and communism. Vinson expressed these views in an address to the West Virginia Mining Institute in December 1914. According to Vinson, "the armed revolution in West Virginia . . . was partially successful because there was no real cooperation" among the coal operators in resisting the UMWA, whom Vinson characterized as robbers and outlaws. Vinson urged the operators present at the meeting to view the union as their enemy and to prepare for war, or face losing the livelihood they gained from mining coal.[14]

One of the operators' associations started in the post-Paint Creek period was the Williamson-Thacker Coal Operators' Association of Mingo County, whose leadership included several of Vinson's political and business allies. Launched in 1913, the association did not take a leading role in the coalfield's affairs until the strike of 1920. Membership in the operators' association was considered a matter of expediency. E. L. Bailey, superintendent of the Portsmouth-Solvay Coal Company, explained that his company only joined the association in order to get the company's production statistics included in a United States Geological Survey study. Throughout the First World War, another organization, the Coal City Club, voiced the concerns of Mingo's coal operators. The quiescence of their men in the 1912–1913 disturbance and their own voluntary institution of workplace reforms inspired a sense of immunity in the Mingo operators; they apparently believed that they had developed a system that diffused the sentiments that led to union agitation.[15]

The Mingo operators were not alone in their undertakings; in fact, their activities were but a pale shadow of their neighbors' efforts. Between 1913 and 1918 throughout southern West Virginia, operators launched the second phase of an anti-union program by adopting a wide array of reforms designed to improve living conditions for their workers while also pacifying them. Many operators voluntarily instituted semi-monthly pay periods and shorter working days. Some brought in welfare workers to teach personal and home hygiene to the miners' wives. They also underwrote the construction of Young Men's Christian Association buildings which provided liquor-free bachelor housing and facilities for leisure activities. In the coalfields dominated by large corporations, a campaign to construct "model company towns" also caught on as companies competed for sober, industrious workers. Most of these reforms reflected the Progressive era belief that an individual's environment heavily influenced his or her behavior. To the operators of southern West Virginia, this meant that if they provided regular pay and a clean, regulated living environment, their workers would have no cause to complain;

A mining camp near Williamson, 1913. *Note the orderly arrangement of the houses. [photograph owned by author]*

any discontent could be attributed to rabble-rousing radical outsiders.[16]

The development of union-resistant operations in Mingo County weakly reflected the development of coalfield giants Logan and McDowell counties. For example, the boom in model company town building only heightened the contrast between Mingo and its neighbors. Although a few of the larger and/or newer operations in Mingo provided their workers with luxuries such as electricity, running water, and movie theaters, none of the company towns in Mingo compared to the planned community atmosphere of Holden, Island Creek's town in Logan County, or to Gary, the flagship company town of McDowell County. Even the companies that provided luxuries like electricity did so at their convenience. Stone Mountain Coal Corporation electrified its company houses, but power was only available at night when it was needed for the preparatory work in the mine. Mrs. Sallie Dickens explained the impact of this when she described how her mother could only iron the family's laundry at or after eleven o'clock at night. As late as 1918, an industry publication, *Coal Age*, praised William Leckie for building a model company town complete with a barbershop at Fireco, Raleigh County. Leckie, who had investments in several southern West Virginia coalfields, did not engage in any comparable improvements at his Mingo County mine.[17]

The uneven and often contradictory actions undertaken by the coal operators in the wake of the Paint Creek-Cabin Creek Strike were paralleled by those taken by the District 17 leadership of the UMWA. The strike and its resolution fatally divided District 17. The men in charge of the Local at the beginning of the strike failed to provide unified, effective direction for the rank-and-file. When the rank-and-file miners put forth their own increasingly radicalized leaders, C. F. "Frank" Keeney, William H. "Bill" Blizzard, and Fred Mooney, the conservative District 17 officials tried to undermine their new rivals. By denouncing these upstarts as Socialists and violent troublemakers, the old line officials were able to hold on to the support of the national organization, despite rank-and-file appeals for their ouster.[18]

After the strike, and feeling secure in the national union's backing, the reactionary leadership of the UMWA's District 17 turned on the individuals who had contributed to the strike's success. The most important victim

District 17 leaders during the Williamson-Thacker Strike. *Left to right: Blizzard, Mooney, Petry, and Keeney. [WVRHC]*

of the post-strike purge was District 17's lead attorney Harold W. Houston. As secretary of the West Virginia Socialist Party, Houston had pressured the party to financially support and publicize the strike. Houston had also ably confronted the operators' attorney Z. T. Vinson during the senate hearings. What the District 17 officials refused to overlook was Houston's alliance with the insurgent leadership.[19]

When Frank Keeney and the other miners had grown frustrated with the weak and vacillating leadership of their elected officers, they had turned to local Socialists for support and direction. Shortly thereafter, the miners initiated a "direct action" campaign and aggressively fought back against the mine guards. The logical inspiration for the rank-and-file's independent tactical turn was Harold Houston. A follower of Big Bill Haywood and a proponent of direct action, Houston, during a speech at Holly Grove tent colony on August 14, had urged the striking miners to follow the strategy Haywood had outlined for the textile workers of Lowell, Massachusetts. Houston had exclaimed, "Now let us divide the work in proportion to the proceeds." District 17's conservative leadership rewarded Houston for first guiding and then defending the miners of

Paint and Cabin Creeks by relieving him of his duties as legal counsel for the union.[20]

The loss of Houston did not deter the insurgent miners' efforts to seize control of District 17. When the national organization refused to unseat the elected leaders, Frank Keeney, Bill Blizzard, Lawrence "Peggy" or "Peg-Leg" Dwyer, and William Petry, bolted District 17 and organized a rival local. For nearly two years the "rump" local, as it was called, submitted dues (which were rejected) to the national union and continued to lobby for recognition as the legitimate leadership of the Kanawha district miners. Finally in 1916, the rival local was dissolved, and a district election was held. Frank Keeney was elected president of District 17 and Fred Mooney was elected secretary-treasurer.[21]

Also in 1916, in what proved to be a fateful decision, District 17 leaders, led by Keeney, broke with Houston and the State Federation of Labor to support Democrat John J. Cornwell for governor. Houston and the State Federationists, who viewed Cornwell as a toady of the B&O and the anti-union capitalists of the state, supported Republican candidate Ira Robinson out of gratitude for his pro-labor stance while on the West Virginia State Supreme Court of Appeals. Cornwell had been advised to court the labor vote in spite of his own conservative disdain for the radical leanings of leaders such as Keeney. Perhaps in retaliation for Hatfield's actions during the last months of the 1912–1913 strike, Keeney and District 17's new leadership responded to Cornwell's half-hearted wooing. However, the relationship between the new officers of District 17 and Cornwell did not long survive the cooperative spirit of the First World War.[22]

Had there been a uniform and uncomplicated application of the coal company reforms, southern West Virginia's operators might have succeeded in lessening the union's appeal to their workforce. Instead, "reform" often created new problems for everyone involved. The operators in the Williamson-Thacker coalfield instituted semi-monthly paydays in 1913. The local press heralded this innovation as a boon to local business because it brought miners to town more often. By contrast, for the operators as a group, semi-monthly paydays were a source of discord and unease. Required to pay their miners twice a month while only bringing in money once a month, small operations strained to secure the necessary capital and resented the larger operations that could bear the expense and always seemed to be waiting for opportunities to eliminate competitors. More frequent paydays also stressed operations because it created more opportunities for criminals, as the case of the Glenalum payroll robbery of 1914 illustrates.[23]

After the Feud and the Massacre, the Glenalum payroll robbery is the third most infamous incident in Mingo County history. During the second week of August 1914, the Glenalum payroll robbery literally shoved the outbreak of war in Europe from local newspaper headlines. In addition to the prurient details associated with murder and banditry, the story of the robbery and the hunt for the thieves also highlights the financial challenges faced by an early-twentieth-century coal company in a remote mining district.

Sometime around 11:00 in the morning on Thursday, August 13, 1914, a Norfolk & Western train delivered to the Glenalum station a $7,000 payroll shipment from the Glenalum Coal Company's headquarters in Lynchburg, Virginia. At approximately 11:05, Glenalum's bookkeeper F. D. Johnson, company doctor W. D. Amick, and electrician Joseph Shielor, left the station on a railroad 'speeder' to travel the four miles down a siding to the Glenalum mine and camp. One and a half miles from the mainline station, a railroad tie across the tracks forced Johnson, Amick, and Shielor to stop the speeder to remove the obstruction. Shots rang out; Amick and Shielor died before their attackers left the scene. Johnson lived long enough to describe what had happened to a traveling salesman who came down the siding tracks shortly after noon. The salesman raced back to the station where notification of the incident was sent to Sheriff Greenway Hatfield.[24]

By mid-afternoon, a posse of sheriff's deputies, Baldwin-Felts detectives, coal company officers and employees, and local citizens descended on Glenalum. Using bloodhounds, the posse tracked the fugitives to a remote mountainous section of Mingo near Ben Creek on the Kentucky border. An early report that the perpetrators of the robbery were natives whose family might harbor them from justice inspired the posse to surround Ben Creek. By Friday, these original suspects were cleared, but by then the posse had closed off all possible escape routes.[25]

Even though the bandits were trapped in Shanty Hollow near Ben Creek, several factors kept the situation from resolution until Sunday morning, August 16. First, torrential rains fell Friday night, temporarily impairing the bloodhounds' ability to follow the scent trail. Second, the bandits proved to be heavily armed; three members of the posse died after coming close enough to the fugitives to exchange gunfire. The final reason for the lengthy manhunt was that the robbers' final hideout in Shanty Hollow was in the heart of "a pathless jungle of dense thickets, deep ravines and a mass of huge vine-covered rocks."[26]

Attempts to assault this position were also complicated by the sheer number of people on the scene. Without the presence of Sheriff Hatfield or an experienced Baldwin-Felts agent, the posse, which had swelled to more than 500 men, turned into a disorganized mob. Throngs of men, women, and children who had followed the posse to the hiding place swarmed in and out of the line of fire. At one point, a burst of gunfire precipitated a stampede of the sightseers who overran the posse, who in turn dared not shoot for fear of striking an innocent bystander.[27]

Sheriff Hatfield arrived on the scene late Saturday as gunfire from the bandits dwindled. Hatfield restored order by commanding the posse to cease firing randomly at the bandits. Someone recommended blasting the fugitives from their hiding place, so sticks of dynamite were brought from the nearby War Eagle coal mine. Throughout the night, posse members lobbed the explosives at the bandits, who at first threw several sticks back at their attackers. The next morning the posse swarmed in and killed the sole survivor of the explosions. Years later members of the posse recounted that the last bandit had jumped up on a log and crowed like a rooster before being shot dead.[28]

The death of the last survivor kept the true identity of the gang members from ever being discovered. However, their physical appearance and a few personal articles found with the bodies suggested that the men were Italians. When the bodies were transported to Williamson, local Italian residents and miners were brought in to see if they could identify the bandits, but no one claimed to recognize any of the dead men. When "one well-known Italian admitted that he would not live long if he divulged any information," local rumors that the dead men were members of the "Black Hand" seemed to be confirmed.[29]

The Glenalum Payroll Robbery. *Mingo County Sheriff Greenway Hatfield and posse and the bodies of the unidentified robbers of the Glenalum Coal Company payroll shipment, August 1914. ["Mingo Republican," 1914, WVRHC]*

Despite the fact that all but $20 of the payroll money was recovered near the bandits' bodies, developments in the weeks after the robbery heightened the unease it caused throughout southern West Virginia. Governor Hatfield had photographs of the bandits' corpses distributed throughout the state, to no avail. The Baldwin-Felts Detective Agency, which had lost two agents in the manhunt, also initiated an investigation that yielded little. The detectives could only confirm that the robbers had arrived in Mingo several days before the robbery; since no one reported their mingling with any local Italians, their identities remained a mystery.[30]

Apart from its local notoriety, the story of the Glenalum payroll robbery also illuminates the complexity and vulnerability of the situation behind the most well-known grievance against the southern West Virginia coal industry. Much has been written about the exploitive nature of paying miners in scrip, but as exposed by the Glenalum incident, by 1914 even small, remotely located companies paid their miners in cash. Common only in the pioneering period, the practice of scrip pay had been made illegal in West Virginia by the first decade of the twentieth century. What persisted in practice and in memory, and was unintentionally misinterpreted in scholarly literature, was the practice of "drawing against pay," in which a miner or his wife could draw scrip that could be used at the company store on credit against his upcoming pay. For many families who became locked in a cycle of debt, scrip may have been the only means of exchange with which they were familiar. However, the Glenalum Coal Company had imported cash to pay its employees, and although the money had traveled safely by rail from Lynchburg, Virginia, the remote four-mile stretch near its final destination provided the opportunity for the crime that led to tragedy.

The other event that encapsulated the 1912–1916 period for the coal industry in Mingo County was the Cinderella mine fire. Opened in 1912, the Sycamore Coal Company's Cinderella mine utilized the latest technology. There were no pick miners at the Cinderella mine; electric machines undercut the coal, and mules and electric motors hauled it to the surface. Three six-foot fans ventilated the mine. Prior to the fire, the only safety concern that had been noted by state mine inspectors was a tendency for the roof to be weak in places, which the company addressed by "careful" timbering and removal of the bad sections. Evan Thomas, the mine's superintendent from its opening through the strike period, was an active community leader. Under his guidance, the operation had become known as one of the model mining towns of the Williamson-Thacker coalfield.[31]

1914, the year of the Cinderella mine fire, was also a peak year for the coalfield before America's entry into World War I. The mines averaged 243 working days in 1914, compared to 227 in 1912 and 226 in 1916. Although the number of companies remained the same between 1912 and 1914 and increased by one by 1916, twelve new mines opened between 1912 and 1914 and three more opened by 1916. The selling price of Williamson-Thacker coal hit a decade-high of $1.17 per ton in 1914. The same ton had brought just $1.01 in 1912 and would drop back to $1.03 in 1916. In 1914, the Cinderella mine ranked sixth out of 21 companies in production, falling behind only the largest producers in the coalfield.[32]

At or near midnight on June 30, 1914, the fanhouse of the well-maintained Cinderella mine caught fire. By the time the night foreman noticed the flames, they had spread to a wooden chute that shunted air from the fans to the newly opened seam, where five men were working. The forced air from the fans propelled the fire into the mine causing the entry timbers, the first stopping, and even the coal in the first room to catch fire. The rapid spread of the fire trapped the five men inside and prevented an easy rescue. For the rest of that night, a crew of men tunneled 27 feet to reach the room where the night crew had been working. By their appearance when found, all five men, including two brothers, had been slowly asphyxiated by gases produced by the burning coal and timbers; they had never attempted to escape.[33]

The *Williamson Daily News* reported that an electrical short circuit had started the fire, but the Department of Mines blamed the miners' deaths on human error. According to State Department of Mines Chief Earl Henry, the mine foreman tried to destroy the chute instead of cutting off the fan. The chief also observed, "Accidents of this character fully demonstrate that fan houses should be constructed of non-combustible material, especially where the fan is operated as a force fan." The Cinderella mine fire demonstrates that in the Progressive Era, even the best maintained mines were dangerous workplaces. The report on the disaster also illustrates that industry experts viewed innovation as a cure for accidents, not a new source of problems; it remained easier to blame humans.[34]

On the eve of America's entry into the World War I, the Williamson-Thacker coalfield stood at a crossroads. Increased production obscured the fact that the field's rate of growth had slowed, while the cost of production had risen. Corporate consolidation in Mingo had continued, but highly capitalized ventures focused more attention on the surrounding coalfields and counties. The Williamson-Thacker coalfield remained a subsidiary interest for these corporate giants. As future events would prove, while not important to the corporate giants in terms of production, the non-union status of the Williamson-Thacker coalfield made it a buffer zone between the unionized Central Competitive Field of Ohio and the Midwest, and the more important coalfields in eastern Kentucky and eastern southern West Virginia.

For the union, if an exploitable situation could be capitalized upon, Mingo would be a potential Achilles' heel and beachhead from which to launch an organization drive. The first West Virginia mine war had failed to resolve the central conflicts in West Virginia's coal industry. The miners of Paint Creek and Cabin Creek had won union recognition and the end of the mine guard system for themselves, but not for their brethren in the surrounding southern West Virginia coalfields. The inability of District 17's conservative leadership to win the strike had resulted in a rank-and-file revolt that brought a new group of officers to power. Young, native, and radical, these men had learned that forging new political alliances increased their bargaining power with state leaders. They also learned that invoking traditional values drew more of their fellow miners to their cause, and that a strategic use of violence increased their access to the court of public opinion.

Observing the forced acquiescence of the Paint Creek and Cabin Creek operators sharpened the desire of anti-union operators to contain the union presence in the state. The southern West Virginia operators organized themselves and instituted social welfare programs and other reforms in an attempt to stave off the union's advance. However, these initiatives met with mixed success in the Williamson-Thacker coalfield. The uneven and half-hearted improvement of living and work conditions, combined with a declining standard of living and job security, increased discontent among Mingo's miners.

The First World War would exacerbate the already strained economic and social tensions in Mingo County and the Williamson-Thacker coalfield. Benefiting less from the war boom and federal regulation of the industry, the Williamson-Thacker coalfield would fall even further behind its neighbors. Efforts to construct social consensus disguised as patriotic "unity" accelerated the polarization of divergent community values in the county. The lingering political grievances from the faction wars, combined with war-induced economic and social destabilization, created in Mingo County the fertile soil the union needed to renew its assault on the anti-union coal empire of southern West Virginia.

When the United States entered World War I, 70 percent of the nation's mechanical energy came from coal, and a full quarter of that coal came from West Virginia. Forty percent of West Virginia's coal came from its four southernmost counties, one of which was Mingo. In 1917, the year President Wilson declared war; coal production in Mingo County's

Williamson-Thacker coalfield reached three million tons for the first time. Traditionally, historians have asserted that the war dramatically increased the demand for coal and helped smaller, peripheral coalfields and mines escape absorption through corporate consolidation. However, after peaking in 1917, Williamson-Thacker coal production declined for over half a decade. The drop in Mingo's production figures before the war's end can only be understood in the context of the Williamson-Thacker coalfield's reaction to the imposition of federal regulation on the coal industry. The wartime "order" imposed on coal production accelerated the marginalization of the peripheral Williamson-Thacker coalfield and reinforced the siege mentality of the men who guided and represented the field's interests. In particular, the coal elite of the Williamson-Thacker coalfield emerged from the war period determined to fight any further interference in their right to autonomously guide the economic future of Mingo County.[35]

On average, the American coal industry experienced "a wonderful development" during the first years of the war. Even Mingo appeared to share in the resurgent prosperity; before the United States' entry into the war, Mingo's existing mines produced more coal in fewer days and profit again began to outstrip the cost of production. Several companies capitalized on the situation by opening new mines, making improvements at existing mines, or selling out to larger corporations. According to the *Mingo Republican,* most of the new operations were opening "almost at the city gates" of Williamson. Relatively new companies that had bought older mines, such as E. L. Sternberger and Stone Mountain, also instituted changes. Sternberger put in a new conveyor belt and made preparations to open a new seam, while Stone Mountain increased the number of its mines by installing two new openings, one west of Matewan and one on Mate Creek, half a mile from town.[36]

The most significant development in the Williamson-Thacker coalfield before the United States entered the war came in February 1917, when the *Mingo Republican* announced that the United Thacker Coal Company (UTCC) holdings were for sale. Encompassing nearly sixty thousand acres and three coal seams of four-, five-, and six-feet thicknesses, United Thacker's lands had never been leased or developed. Given the size and potential value of the UTCC lands, only the most heavily capitalized corporation could have afforded to buy them outright. To ensure its potential investment, U.S. Steel sent Colonel Edward O'Toole, the general manager of its southern West Virginia holdings, and Howard N. Eavenson, the leading coal expert in America to "inspect" the UTCC lands. In marked contrast to the speculative gambles of a generation before, the O'Toole/ Eavenson inspection involved the shipping of nine carloads of coal to laboratories for analysis. Although the U.S. Steel-United Thacker deal was not finalized for months, the prospect of the giant corporation's move into Mingo dovetailed with existing improvement and expansion efforts to advance the price of coal lands in the county.[37]

Six months after the *Mingo Republican* first reported the possible sale of the United Thacker Coal Company's land in Mingo and Logan counties, the U.S. Coal & Coke Company bought fifty thousand acres, the majority of UTCC's Mingo holdings. The purchase included "all of UTCC holdings on Island Creek in Logan County and on Gilbert, Horsepen, Ben, and Beech creeks in Mingo and part of its holdings on Pigeon Creek." The *Republican* celebrated the news because U.S. Coal & Coke would have to expand rail development in order to fully exploit its property and because several impediments had been overcome to secure the sale.[38]

Mingo County's experiences in the 1914 to early 1917 period reflected similar developments in southern West Virginia's other coal counties. W. R. Thurmond, a pioneering coal operator and amateur historian of southern West Virginia's coal history, observed that "the runaway market during the First World War attracted a number of operators that soon disappeared" leaving "the very large and heavily capitalized companies . . . to dominate the market."[39]

The creation and short history of the Wigarb Mining Company illustrate how the boom of the early war years affected coal production in Mingo County. The chief investors and officers of Wigarb included L. E. Armentrout, the superintendent of the Borderland Coal Corporation, and George Bausewine, Jr. and C. M. Gates of the Crystal Block Coal & Coke Company operations at Sprigg. Armentrout ran Borderland for several powerful politicians and businessmen from Virginia, and his importance to the local operation was signified by the designation of the local post office as "Armen." As the chief officers of the U.S. Coal & Coke (U.S. Steel) subsidiary Crystal Block mines at Sprigg, Bausewine and Gates were equally powerful figures in the local coal community. Both were member-officers of the Coal City Club. Bausewine also served as an officer of the Williamson Coal Operators' Association. The organization and demise of Wigarb reveal the go-go atmosphere of the early war period, when the lure of profit inspired the local managers of larger corporations to sink their own funds into coal production. By 1920, or just over two years later, Wigarb was producing no coal, and federal wartime policies and the post-war recession had negated the experience of its organizers.[40]

The United States' entry into World War I significantly affected relations both within the coal industry and between the industry's leaders and government officials. Disillusioned by previous peacetime efforts to bring under control the coal industry's endemic and persistent problems with cutthroat competition and overproduction, Progressives in the Wilson administration used the war crisis to seize control of the industry. Initially justified as an attempt to punish and prevent further war profiteering by the corporate giants who dominated the national coal industry, federal regulation ultimately became another weapon of those same corporations in their fight to eradicate their smaller, poorly capitalized competitors. This phenomenon is illustrated by the contrasting response to federal regulation by the Williamson-Thacker coal elite and the more powerful coal men of the Guyan and smokeless coalfields.[41]

In late June 1917, and only one week following the publication of a Federal Trade Commission report advocating the "pooling" of all coal production, 400 operators converged on Washington. During a three-day conference sponsored by the Committee on Coal Production, FTC Commissioner Frank J. Fort urged the operators to lower prices voluntarily. The coal industry followed Fort's advice and set maximum prices at three dollars a ton. The plan backfired miserably; the voluntary maximum ignited public outrage. The cooperative action among so many coal operators also convinced many Progressives in the Wilson administration that a "coal trust" existed and needed to be policed.[42]

The industry's second move to protect its interests only reinforced the belief that the coal operators' actions stemmed from ulterior motives. Between late July and early August 1917 the coal industry founded the National Coal

Association. Previous attempts to create an effective national organization that could promote a good public image for the industry and protect its interests in the halls of power had failed. The last serious effort, in 1909, had collapsed because southern coal operators refused to join out of fear that membership translated into tacit recognition of the union. The successful organization of the NCA in the summer of 1917 proved to be a strategic blunder. The NCA's apparent unity led government officials to condemn "an essentially fragmented and leaderless industry as if it were a monopoly."[43]

As Allen Batteau points out, "When America entered the war, the [coal] industry . . . was disorganized, highly competitive, inefficient and wasteful." Once heralded as the very source of America's modernization, the national coal industry was perceived as a reactionary, ill-managed relic of a bygone age by the height of the Progressive era. Citing the need to ensure a steady supply of coal for the war effort, President Woodrow Wilson authorized the federal government to control the national distribution of coal on August 23, 1917. The president's directive divided the country into 29 production-distribution districts. Particularly angered by the coal industry's voluntary maximum price of $3.00 a ton, President Wilson used the June FTC report on the cost of coal production to set maximum prices at nearly a third less. West Virginia's coal prices were set at $2.00 a ton for run-of-mine, $2.25 for prepared sizes, and $1.90 for slack or screened coal. Following Wilson's lead, Congress passed the Lever Act, which authorized the creation of an agency to implement regulatory policies regarding the sale and distribution of coal. Stunned by the public and federal reaction to what it believed had been a patriotic action, the coal industry felt betrayed by the national government's rejection of its voluntary self-regulation. Although the price control recommendations had been based on FTC estimates of production costs, the move still outraged the industry and its political allies.[44]

The coal industry had so little clout in Washington it could not even influence the wording of the Lever Act, which passed through the Senate in a vote of 50 to 27. The lack of unity and the level of disorganization were so high in the industry that the newly founded National Coal Association could not even orchestrate an official response to federal regulation. Some operators wanted to challenge Wilson's prices in the courts, but others,

Wells Goodykoontz, Williamson Republican and sometimes Hatfield ally.
Goodykoontz was serving a term as U.S. Congressman when the Matewan Massacre occurred. [WVRHC]

like James D. Francis, president of Island Creek Coal Company, believed that reasonable profits could be earned even with the $2.00 maximum price. The division within the National Coal Association reflected the differences between operations like U.S. Coal & Coke, Island Creek, and the myriad of smaller companies that dotted southern West Virginia.[45]

Companies such as Logan County's Island Creek and McDowell County's U.S. Steel subsidiary U.S. Coal & Coke, which were securely capitalized and integrally linked to multi-industry conglomerates, could absorb the cost of federal regulation. Moreover, these leading producers of coal ultimately were able to emulate better organized industries with political influence and establish a successful working relationship with the Wilson administration. Comparison of U.S. Steel's 1914 and 1917 profit figures illustrates how the large corporations not only survived the vicissitudes of the wartime economy but in fact thrived under government regulation. In 1914, U.S. Steel's annual profits were $76 million; in 1917 annual profits were $478 million.[46]

By contrast, even before the imposition of federal regulation, the representatives of the smaller, peripheral coalfields angrily denounced the move. In June 1917, the president of West Virginia's state senate and Mingo's own state senator Wells Goodykoontz wrote to two West Virginia congressmen, Senator Howard Sutherland and Fifth District Representative Edward Cooper, condemning national control of coal production as a "fanatic extreme." Goodykoontz asserted that the adequate production of coal could only be secure if businessmen were "left free to earn reasonable profits." Goodykoontz declared that the real trouble lay with the railroad industry. If the railroads would deliver on their promises of more coal cars and build more sidings to mine sites, coal prices would come down without "the county [being] Germanized."[47]

Throughout the course of the war, Goodykoontz maintained his barrage on behalf of small coal producers. In addition to berating the railroad industry, he railed at the injustice of the escape of steel, copper, and cotton from any regulation. As Goodykoontz noted, even wheat had been protected by a minimum price; only coal had been singled out "as a burnt offering to the people." The coal boom of 1916 had not only elevated the price of coal, as Goodykoontz noted; the increase in the expense of coal production

A Mingo County Truck Mine. *Mingo's truck mines were the hardest hit by World War I and were the sources of the UMWA's greatest organizing success in 1920. [WVRHC]*

when combined with federal price-fixing threatened to "whittle down coal prices to a ruinous figure." The anger and fear Goodykoontz was expressing was grounded in what was proven to be true. Unlike the large, highly capitalized coal producers, such as U.S. Steel, small and/or poorly or precariously capitalized companies simply would not be able to absorb the higher production costs, competition for workers, and the limited rate of return that accompanied wartime regulation.[48]

Created by the Lever Act in late August 1917, the Fuel Administration directed the implementation of the guidelines stipulated by President Wilson's executive order for the duration of the war. The head of the agency, Dr. Harry A. Garfield, and his top assistants were almost all professional academics with no intimate knowledge of the industry they intended to supervise. The son of President James A. Garfield and a personal friend of Woodrow Wilson, Harry Garfield had been the president of Williams College before accepting appointment as fuel administrator. His sole preparation for the position had been a previous assignment to study the impact of federal price control on wheat production. Although many in the coal industry initially applauded Garfield's ascension as fuel administrator because they feared the appointment of more radical candidates,

Garfield's attitude toward the business community soon elicited concern. When given control of the Fuel Administration, Garfield and his staff excluded representatives of the coal industry from critical decisions.[49]

Early support for Garfield disintegrated into bitter denunciation. The *Mingo Republican* and self-appointed Williamson-Thacker spokesman Wells Goodykoontz were particularly angered by Garfield and staff's ignorance or lack of concern for the industrial idiosyncrasies that affected coal mining along the Tug Fork River. The efforts of the newspaper and Senator Goodykoontz highlight how the macrocosmically-designed Fuel Administration policies disadvantaged the operators of the Williamson-Thacker coalfield. Among the smallest and weakest mines in southern West Virginia were the "wagon mines" that did not produce enough coal to either justify or pay for a direct rail connection or "siding" to their mines. Instead, these mines had to haul their coal in wagons to the nearest coal-loading rail location, hence the name "wagon mines." On behalf of the wagon mines in Mingo County, Goodykoontz launched a campaign to persuade the Fuel Administration to make exceptions for these small mines.[50]

Goodykoontz began by seeking clarification of the Administration's policy regarding wagon mines. Responding for the Fuel Administration, P. B. Noyes explained to Goodykoontz that reasonable charges for hauling could be added if the mines were transporting coal to local consumers or to the railroad if the railroad was purchasing coal for its own consumption. An article in the September 27, 1917, issue of the *Mingo Republican* revealed that these stipulations provided no relief for Mingo's wagon mines. Convened to discuss ways and means of securing relief from the price of $2.00 per ton for coal, a meeting of Mingo's wagon mine operators resulted in the decision to send a three-man committee to Washington to meet with Fuel Administrator Garfield. The impetus for the meeting and the appeal to Garfield stemmed from the fact that many of the small operators in Mingo had been compelled to stop producing coal.[51]

By the time Mingo's small operators decided to appeal for relief, the Fuel Administration had been forced to acknowledge that coal production had fallen following the president's announcement. When Garfield realized that the imposition of the $2.00 limit had also priced out the marginal producers, he moved to rectify the situation. Just two weeks after the meeting in Williamson, the Fuel Administration authorized wagon mines to charge an additional 75 cents per ton where the coal is loaded into box cars. This small concession did not help the mines already driven out of the market, nor did it provide relief for the wagon mines of Mingo County. As the *Mingo Republican* revealed, unlike other railroads, the N&W did not use many box cars for transportation, and coal had to compete with merchandise and other freights for the remaining limited space.[52]

The actions of the Norfolk & Western in the late summer and early fall of 1917 further exacerbated the frustration felt by Mingo's small operators and their advocate Goodykoontz. While price regulation cut its first swathe through Mingo's small mines in August and September 1917, the N&W purchased operations at Vulcan and Chattaroy in order to secure a coal supply in Mingo County. The N&W's expansion of its captive coal supply nullified the minor concession of the Fuel Administration which allowed wagon mines to charge extra for coal provided to railroads. Moreover, the N&W's insufficient supply of freight cars further compounded the wagon operators' difficulties. Obsessed with holding coal operators accountable for coal price inflation, the federal government blithely disregarded the N&W's self-serving actions.[53]

The Fuel Administration also ignored the economic implications of its policies based on geographic divisions it had assigned. For example, West Virginia and Kentucky coal sale rates were set differently despite the irrelevance of their shared border. Several mines in the Pond Creek district of Pike County, Kentucky, actually dumped their coal onto trains passing through the Williamson-Thacker district of Mingo County, West Virginia. The mines of both the Pond Creek and Williamson-Thacker districts tapped into the same three seams of coal: Thacker, Pond Creek, and Winifrede. Disregarding the unique conveyance system and the commonality of the coal produced in the two districts, the Fuel Administration allowed the Pond Creek operators to benefit from the Kentucky price for coal, which exceeded the West Virginia rate by 40 cents per ton.[54]

Although the challenges faced by operators in coalfields like the Williamson-Thacker coalfield failed to substantively affect Fuel Administration policies, the onset of the winter of 1917–1918 dramatically influenced the course of government control of the national coal industry.

A difficult adjustment to wartime industrial production and one of the coldest winters in American history threatened social stability and led to outbreaks of civil unrest. Residents of Ohio communities suffering from coal shortages tore up train tracks bypassing them on the way to industrial centers, while people in Brooklyn rioted for fuel for their homes. Allowed by the government to be "underfinanced and under maintained" for years, the railroad industry could not meet the demands of the war and the public that winter. By December, the railroads were effectively paralyzed.[55]

President Wilson and other federal officials were forced to acknowledge that the coal industry was not solely responsible for the inadequate and uneven distribution of the nation's fuel supply. However, rather than lift price controls on coal and stimulate production, on December 26, 1917, President Wilson "abruptly placed control of the [rail] industry in the hands of his son-in-law, Secretary of the Treasury William Gibbs McAdoo." Unlike the coal industry, the rail industry's full property rights were assured, and the railroads were promised annual compensation equal to the average of the three years prior to 1917. Many railroads did not benefit from these pledges because they had been in an earnings slump since 1915. However, the N&W, which serviced Mingo County and the Williamson-Thacker coalfield, was primarily a coal carrier that had escaped the decline in revenue and had its compensation set at $20 million. The income security proffered to the Norfolk & Western contrasted sharply with the disregard exhibited toward the idiosyncratic needs of the Mingo coal operators. The exacerbation of those needs by the railroad's limitations further underscored the lack of balance in federal policy.[56]

Nationally, the creation of the Railroad Administration and the conciliatory gestures of the Fuel Administration helped stem the fuel crisis caused by the unusually severe winter. Coordinated by the two agencies, coal production increased as a result of the ending of cross-hauls and an improved car supply. Resolution of the immediate impediments to sufficient transportation allowed the Fuel Administration to turn to appeasing public outrage and fears.[57]

Alarmed by the public unrest caused by coal shortages, the Fuel Administration also authorized the expansion of its investigatory operations at state and local levels. State Fuel Administrators, appointed "to help insure equitable local distribution of coal," were supposed to be public-minded civic leaders with no connections to the industry. West Virginia's state fuel administrator, J. Walter Barnes of Fairmont, met Dr. Garfield's criteria. Barnes' professional credentials bore a distinct Progressive stamp. A lawyer and educator, Barnes had been principal of the Fairmont Normal School and the general manager of a utility company. At the time of his appointment, he was a member of Fairmont's city commission and chair-man of the Marion County Council of Defense.[58]

State Fuel Administrator Barnes appointed former State Senator Hiram S. White of Matewan as Mingo's local fuel administrator. White selected William T. Meade of Kermit and Edward R. Juhling of Williamson as his assistants. White, Meade, and Juhling were empowered to investigate re-ports of operators and/or retailers charging "greater prices than allowed" under the federal mandate. Within three weeks of their appointment at the beginning of January 1918, Mingo's fuel administrators had heard sev-eral such cases but refused to publicly reveal any details or to say whether any of the accused was guilty. The Board simply reported that no action was recommended.[59]

Despite the official silence on who faced charges before Mingo's Fuel Administration board, the membership of the board itself reveals that the application of Fuel Administration policies could not have escaped the va-garies of local politics. White, Meade, and Juhling were all veterans of Mingo's political intrigues and faction wars, and all three were known "bolters." The older brother of Dr. Israel C. White, West Virginia's state ge-ologist, Hiram S. White had frequently benefited from patronage. Before the war, his wife served as the postmistress of Matewan while he acted as a state workmen's compensation investigator. A resident of Matewan since the county's founding, White had also actively participated in the county Republican Party's bitter factional fights. The sole Democrat on Mingo's Fuel Administration board, William T. Meade, led the Democrats who controlled Kermit. Edward R. Juhling, a founding member of the Coal City Club, had repeatedly protested Norfolk & Western and federal policies deemed unjust and injurious to local business.[60]

One example of charges against a local coal company underscores both the means of evading federal price policy and the role of local politics in accusations of abuse. Details of the alleged conspiracy also illuminate the economic hierarchy of southern West Virginia's coal companies. According to the charges, a well-connected large corporation's subsidiary, based in Charleston, used the officers of a local entrepreneur-backed company to prey on financially strapped small mines. Coincidentally, the accuser and the accused shared a long but recently heated history of political animosity.

In a letter to the assistant U.S. district attorney for southern West Virginia, Williamson attorney S. D. Stokes accused J. Levine, the general manager of the Levine-Goodman Coal Company, Blaine York, a local attorney, and the Logan Pocahontas Fuel Company of violating the federal regulations on the price of coal. According to Stokes, Logan Pocahontas deposited money in Williamson that York and Levine used to pay "bonuses" for coal from small mines. Stokes learned of the scheme from Levine-Goodman's president, who professed ignorance until Levine offered to divide the profit with him. Assistant district attorney Kelly informed Stokes that a special agent would be assigned to the case but also warned that the investigation could be delayed by several months.[61]

Stokes, Levine, and York were old adversaries in Williamson municipal and county-wide politics. Stokes served as a lieutenant in the Democratic City Ring controlled by Wallace J. Williamson. Also from Williamson, Levine and York were Republicans with close ties to the Hatfield machine, and both had been elevated to power in city politics when Henry D. Hatfield orchestrated the revocation of Williamson's charter which had stripped the City Ring of its power. Stokes had spearheaded the effort to undo the charter, even going so far as to seek advice from John W. Davis, then serving as assistant United States solicitor general.[62]

Whether Stokes had first brought charges against Levine and York before Mingo's Fuel Administration board is unknown. However, the tone of his appeal to Assistant District Attorney Kelly reveals that he did expect satisfaction from the federal officer. As if to emphasize the probability of York's guilt, Stokes reminded Kelly that the district attorney's office had investigated York for another unrelated federal violation.[63]

The detail not mentioned in the Stokes-Kelly correspondence was Kelly's familiarity with Mingo County politics. For four years Kelly had helped prosecute Mingo's "Election Fraud" cases. The primary defendant in those cases was Greenway Hatfield, chairman of Mingo's Republican Party, brother of Governor Henry D. Hatfield, and the arch rival of Stokes's mentor Wallace J. Williamson. The final "Election Fraud" case had ended in acquittal only three days before Stokes wrote to Kelly. By reporting the Levine-York incident, Stokes offered Kelly a new opportunity to strike at Hatfield through his associates, which in turn would allow Mingo's Democrats to accuse the county's Republican businessmen of being unpatriotic profiteers. Kelly carefully hedged around Stokes' implicit offer. He explained why York had not been prosecuted on the other matter and then closed the letter with a carefully phrased, "you can rest assured, that the matter will be attended to."[64]

Only one day after the *Mingo Republican* first reported on the activities of the local Fuel Administration, Fuel Administrator Garfield and Rail Administrator McAdoo reconfigured federal regulation of the production and distribution of coal for the second and last time. The Garfield-McAdoo "Zone Plan" divided the country into 20 consumption districts and 11 production districts. Set to take effect on April 1, 1918, the Zone Plan revived Garfield's reputation with the coal industry, which hailed the plan as finally making "the distribution of coal rational and efficient." To the operators of the Williamson-Thacker coalfield of Mingo County however, the Zone Plan reinforced their feeling of being ignored and discounted in favor of other and/or larger coalfields.[65]

From the macrocosmic perspective, however, the plan made sense. First, the plan required consumers from the Midwest to purchase coal from their region instead of from Pennsylvania, West Virginia, or eastern Kentucky. Second, the Norfolk & Western and the coal producers along its line were confined to distribution zones allocated to the "Pocahontas zone." Requiring Midwestern consumers to purchase the coal mined nearest them alleviated congestion on the rail lines. In turn, the policy opened "bottlenecks in the Appalachian gateways" and allowed 9 percent more coal to move through the mountains. Under the Zone Plan, production and prices rose, and the operators could now sell all their output at prices nearly double those of a few years before.[66]

From the microcosmic perspective of Mingo's operators, the same two elements of the plan seemed designed to undermine the marketability of

Mingo's coal. By the late teens the Midwestern or Lake Cargo trade was the only market where Williamson-Thacker coal could compete against both the Central Competitive and smokeless coalfields. As a non-union coalfield, despite the rate differentials on coal shipments, Williamson-Thacker coal sold more cheaply than the Midwestern coal produced in the unionized mines of the Central Competitive Field. It also underpriced the Pocahontas coal shipped west because the Pocahontas coal producers paid an even higher rate differential than the Williamson-Thacker operators. Deprived of the Midwestern outlet, Williamson-Thacker coal could only be shipped to areas long dominated by the smokeless coalfields' producers of superior quality coal. To the operators of the Williamson-Thacker coalfield, the Zone Plan enforced yet another squeeze while the benefits accrued to larger, more powerful interests.[67]

Relations between the coal companies, Mingo's county government, and private citizens suffered during the spring and summer of 1918. Cutthroat competition also intensified the county's traditional adversarial politics between Democratic and Republican businessmen. Williamson Democrat S. D. Stokes's charges of profiteering against the Republican-backed Levine-Goodman Coal Company were leveled one month after the Zone Plan went into effect. Other companies directed their cost-saving efforts to Mingo's county government. The newly renamed Allburn Coal Corporation challenged a $137.15 tax assessment because its West Virginia tax liability was limited to a shared responsibility for two acres of land, a tipple, and personal property that consisted of "two old boilers and a little machinery." After pleading its case before Mingo's Board of Equalization, Allburn had to wait four months before receiving the ruling that it had been taxed erroneously.[68]

Despite the work stoppage caused by the Influenza Epidemic in September and October of 1918, American coal production and consumption reached all-time highs in 1918. When the Armistice took effect on November 11, the national surplus of nearly one hundred million tons of coal confirmed the miners' patriotic effort. However, as historian James P. Johnson commented:

As whistles blew on mine tipples and individual plants, buyers of war materials jammed switchboards with order cancellations.

> Within a month the War Department retracted half its outstand-
> ing contracts . . . [and the] dramatically reduced demand for coal
> brought falling prices, particularly for poorer grades.

Moreover, the Armistice technically did not end the war or wartime
regulation of the coal and rail industries. The dismantling of federal control
dominated the year following the cessation of hostilities. The February 1,
1919, issue of the *Williamson Daily News* gleefully announced that "All
Control Exercised by the Fuel Administration Is Over." Expressing the
hopes of Mingo's operators, the newspaper opined that the end of regulation
would reinvigorate coal production in the county. As the events of the rest
of 1919 revealed, the *Daily News*'s announcement proved premature.[69]

The general recession that followed the end of hostilities persisted into
the first quarter of 1919. Even though the coal industry had not produced at
maximum capacity during the war, the drop in demand created a 100 million ton surplus. This reservoir of excess coal inspired Fuel Administrator
Garfield's decision to suspend but not rescind the agency's regulatory stipulations regarding production zones and prices.[70]

Within months, Mingo's leaders realized that the lifting of certain Fuel
Administration policies and the suspension of some wartime regulations
did not mean that all federal interference in the coal trade had ended. In
May, now-U.S. Congressman Wells Goodykoontz released an angry letter
to the *Williamson Daily News* in which he protested the actions of Henry
Spencer, a regional purchasing director of the Railroad Administration.
According to Goodykoontz, Spencer had ordered operators in northern
Ohio, Michigan, and Indiana to purchase only Ohio coal. Considering that
at the beginning of the war nearly half of Mingo's coal output traveled to
the now restricted area, one can understand Goodykoontz's frustration.[71]

However, by the time of the publication of Goodykoontz's letter, market
forces had already swung favorably. In the second quarter of 1919 "a sharp
upswing of business activity and employment" ended the post-Armistice recession. Barely a week after showcasing Goodykoontz's letter, the *Daily News*
reported that the venerable Thacker Coal & Coke Company, which had been
shut down since the suspension of federal regulation, was reopening.[72]

As spring warmed into summer, other indicators of revival dotted the pages of the *Williamson Daily News*. A single issue of the newspaper carried two notices that reveal how work picked up in the Williamson-Thacker coalfield. The Sharondale Coal & Coke Company published a notice that stated simply "work has been resumed" and told where miners and company men could apply for employment. The West Virginia Bi-Product Company advertised for twenty-five coal loaders, noting that it had enough orders to run full-time and could offer loaders and laborers steady work at "good wages."[73]

Tables 9–11 at the end of this chapter reveal the effects that war and government regulation had on Mingo and its two powerful neighbors, Logan and McDowell. As the tables show, federal regulation affected the coal industry in Mingo and Logan Counties similarly. The number of mines and miners in Mingo and Logan counties increased, and the mines in both counties also worked more days during the war. Only McDowell County suffered more losses than gains in the categories that indicate growth.

What the numbers and percentages do not convey is the way federal regulatory policy exacerbated the crisis mentality of Mingo's coal operators. Local coal industry advocates resented the Fuel Administration's failure to recognize and accommodate the unique position of the Williamson-Thacker coalfield. The imposition of the Zone Plan compounded the injustices Mingo County's coal elites believed had been heaped upon them because the distribution system denied access to the Midwestern coal market, the only arena where the Williamson-Thacker coalfield had enjoyed an advantage over its larger neighbors.

One statistic underscores the scale of Mingo's loss. Just as Island Creek president James D. Francis had predicted, federal regulation benefited Logan County, where coal production increased by nearly a million tons during the war. By contrast, tonnage output in Mingo and McDowell Counties fell. But McDowell, despite suffering an 800 percent greater loss, still produced almost six times the amount of coal mined in Mingo. As a result, the operators in Mingo County emerged from the war more determined than ever to resist any constraints on their autonomy.[74]

In 1920, when U.S. Department of Labor mediators followed union organizers into Mingo County, they were firmly rebuffed. The operators of the Williamson-Thacker coalfield would brook no more interference in the running of their mines, not even from the United States government. What they could not know was that they would not, could not win the conflict that began in 1920. It was not the United Mine Workers of America who would defeat the Williamson-Thacker operators; it was their own allies. Within five years of the end of the Williamson-Thacker strike, large corporations would finally absorb this troublesome periphery and eliminate, or so they thought, any threat to their control of southern West Virginia's coal.[75]

Production Figures	1917	1918	Change
Number of Mines	53	58	up 9%
Tonnage Output	3,207,162	3,104,419	down 3%
Total Number of Employees	3,383	3,581	up 5%
Number of Pick Miners	182	194	up 6%
Number of Machine Miners	1,132	1,064	down 4%
Miners as a Percentage of Total Employees	39%	35%	down 4%
Number of Days Worked	220	238	up 8%
Average Price per ton	$1.84	$2.77	up 34%

Table 9. The Impact of Federal Control on the Coal Industry in Mingo County, 1917–1918* Source: "Annual Reports of the West Virginia Bureau of Mines, 1917–1918." *Mingo County was in the Williamson-Thacker coalfield.

Production Figures	1917	1918	Change
Number of Mines	124	141	up 9%
Tonnage Output	8,859,122	9,229,975	up 4%
Total Number of Employees	7,965	9,653	up 17%
Number of Pick Miners	203	634	up 68%
Number of Machine Miners	3,591	3,450	down 4%
Miners as % of Total Employees	48 %	43%	down 5%
Number of Days Worked	189	203	up 7%
Average Price per ton	$2.06	$2.83	up 27%

Table 10. The Impact of Federal Control of the Coal Industry in Logan County, 1917–1918* Source: "Annual Reports of the West Virginia Bureau of Mines, 1917–1918." *Logan County was located in the Guyan coalfield

Production Figure	1917	1918	Change
Number of Mines	152	157	up 3%
Tonnage Output	18,671,942	17,812,416	down 5%
Total Number of Employees	19,170	17,639	down 8%
Number of Pick Miners	4,195	3,288	down 22%
Number of Machine Miners	4,801	4,059	down 15%
Miners as % of Total Employees	47%	42%	down 5%
Number of Days Worked	225	250	up 9%
Average Price per ton	$2.01	$2.85	up 29%

Table 11. The Impact of Federal Control on the Coal Industry in McDowell County, 1917–1918* *Source: "Annual Reports of the West Virginia Bureau of Mines, 1917–1918."*
*McDowell was located in the Pocahontas coalfield.

NOTES

1   Harold (Howard) Radford interview with John Hennen, Summer 1989 Matewan Oral History Project.

2   See author's dissertation chapter six for explication with specific statistics.

3   "Houston Brief," 46; *Annual Reports of the West Virginia Department of Mines*, 1902–1916; Chernow, 101, 146–147.

4   Thomas H. Claggett to W. Come, general manager, Pocahontas Coal & Coke Company, September 16, 1913. Pocahontas Land Corporation Collection, ERCA; Until the labor strife of the 1920–1922 period, the Williamson-Thacker field included the Pond Creek area of Pike County, Kentucky. During the strike, the operators along Pond Creek sought and were granted an injunction against the UMWA. Lane, 71–72; The company that became Island Creek in 1915 began purchasing land in the Pond Creek field in late 1911. *Coal Trade Journal* 1 (December 1911): 262; Island Creek eventually expanded into Mingo County in the 1940s when it purchased the mines and holdings of Red Jacket Consolidated Coal & Coke.

5   Corbin, *Life, Work and Rebellion*, 87.

6   Corbin, *Life, Work, and Rebellion*, 87–88; Lee, 21.

7   Corbin, *Life, Work, and Rebellion*, 88–89.

8   Corbin, *Life, Work, and Rebellion*, 89, 90, 95.

9 Corbin, *Life, Work, and Rebellion*, 95–97; Harold W. Houston backed Robinson, but other leading West Virginia socialists backed Lilly; Barkey, 188.

10 Corbin, *Life, Work, and Rebellion*, 97.

11 Corbin, *Life, Work, and Rebellion*, 98, 98–99.

12 Corbin, *Life, Work, and Rebellion*, 99.

13 Corbin, *Life, Work, and Rebellion*, 113, 110.

14 A full-scale propaganda campaign, the third element of the coal operators' anti-union initiative, was not used wholesale until the Williamson-Thacker strike began in 1920. McIntosh Memoir, WVRHC; Roy B. Naylor made the "in union" observation in an article titled "Commercial Organizations in West Virginia," published in James M. Callahan's *Semi-Centennial History of West Virginia* (Charleston, WV: n.p., 1913): 319–322; In an article first published in the *Pittsburgh Gazette*, and reprinted in the *Williamson Daily News*, the founding of the West Virginia Coal Operators Association was described as "the cost of crooked politics." Citing attacks on business by aspiring politicians and recent acts of the West Virginia legislature, the West Virginia Coal Operators' Association presented itself as "an organization for self-protection"; *Williamson Daily News*, July 24, 1915; Z. T. Vinson, "Advocating Co-operation and Organization of West Virginia's Coal Operation," an address before the West Virginia Mining Institute, December 10, 1914, in Huntington, WV (n.p.n.d.), 5–6.

15 Corbin, *Life, Work, and Rebellion*, 113; *Williamson Daily News*, June 29, 1919; In press statements and other public documents, the formal name of the Williamson-Thacker field's operators' association, "The Operators' Association of the Williamson-Thacker Coalfield" was always used; "Testimony of E. L. Bailey, superintendent of the Portsmouth-Solvay Coal Company" *West Virginia Coalfields*, 288; Source for reason for joining operators' association: *Williamson Daily News*, January 4, 1919; Source for Bailey as owner-operator: *Mingo Republican*, March 15, 1917.

16 Corbin, *Life, Work, and Rebellion*, 120, 125,117; Diner, 201; Robert F. Munn, "The Development of Model Company Towns in the Bituminous Coalfields," *West Virginia History* 40 (Spring 1979): 245.

17 Munn, 248. Munn identifies Holden as one of the first and most well-known model towns in the bituminous coalfields; Sallie Dickens interview with Rebecca J. Bailey, Summer 1990 Matewan Oral History Project; "Fireco: A New Mining Town in West Virginia," *Coal Age* 12 (September 12, 1918): 1217.

18   Corbin, *Life, Work, and Rebellion*, 100.

19   Corbin, *Life, Work, and Rebellion*, 100. Corbin quotes an appeal sent to International headquarters in which the old District 17 leaders were described as "conservative, corrupt, and undemocratic"; "Telegram from Harold W. Houston to the National Executive Committee, Socialist Party, read at the meeting of the committee, May 11, 1913," reported by Grace Silver in "National Committee Meeting, Socialist Party, in *International Socialist Review*, Pamphlet 191, WVRHC; The assertion about District 17 officials is based on Houston's resumption of his role as District 17 attorney when Keeney and the other insurgents legally took control of the local.

20   Corbin, *Life, Work, and Rebellion*, 89; Barkey, 114; "Extracts from Speech of Harold W. Houston at Holly Grove, on Paint Creek, August 14, 1912," in *Conditions in the Paint Creek District, West Virginia: Hearings Before a Subcommittee of the Committee on Education and Labor*, Part 3 (Washington: GPO, 1913): 2263; *Wheeling Intelligencer*, March 20, 1914.

21   Corbin, *Life, Work, and Rebellion*, 100.

22   Barkey, 189; Fred Mooney, *Struggle in the Coalfields: The Autobiography of Fred Mooney*. (Morgantown, WV: West Virginia University Library, 1967), 59–69; Fisher, 380.

23   *Mingo Republican*, August 8, 1913; Hinrichs, 40–41.

24   *Mingo Republican*, August 13, 1914 and John A. Velke, *Baldwin-Felts Detectives, Inc.* (Richmond, VA: s.n., 1997), 137–141.

25   *Mingo Republican*, August 13, 1914, and August 20, 1914.

26   Emick R. Walls, "West Virginia's Greatest Manhunt," *West Virginia Illustrated*, 2 (May–June 1971): 29–33.

27   Walls, 29, 33; *Mingo Republican*, August 20, 1914.

28   Walls, 32; *Mingo Republican*, August 20, 1914; Harold (Howard) Radford interview.

29   The robbers had also initially been suspected of being African American. Johnny Fullen interview with Rebecca J. Bailey, Summer 1990 Matewan Oral History Project; *Mingo Republican*, August 20, 1914.

30   *Mingo Republican*, April 8, 1915; *Mingo Republican*, August 20, 1914; Walls, 29; Velke, 140–141.

31   *1912 Annual Report of the West Virginia Department of Mines*, 368, 65, 315; State mine inspectors filed virtually identical positive reports over the next sever-

al years. The Cinderella mine rated "good" or "very good" in all categories includ-
ing drainage, ventilation, and lack of explosive gas; Source for the fact that the
Cinderella mine did not employ pick miners: *Mingo Republican*, March 9, 1916;
Thomas served as president of the Lee district board of education.

32   Based on data extrapolated from *Annual Reports of the West Virginia
Department of Mines*, 1912–1916; Based on data extrapolated from the *1914 Annual
Report of the West Virginia Department of Mines*, production tables, 146–147.

33   "Mine Fire: Cinderella Mine of the Sycamore Coal Company," in the *1914
Annual Report of the West Virginia Department of Mines*, 70.

34   *Williamson Daily News*, July 9, 1914; *1914 Annual Report of the West Virginia
Department of Mines*, 70; Dillon, 47–49.

35   Corbin, *Life, Work, and Rebellion*, 185; "Houston Brief," 40; *1917 Annual
Report of West Virginia's Department of Mines*; Soule and Carosso, 507; *1917–1922
Annual Reports of West Virginia's Department of Mines*.

36   *Mingo Republican*, February 15, 1917; *Mingo Republican*, March 15, 1917.

37   *Mingo Republican*, February 15, 1917; *Mingo Republican*, March 15, 1917.

38   *Mingo Republican*, September 27, 1917.

39   Thurmond, 40.

40   *Mingo Republican*, May 3, 1917; For more on Borderland, see Crandall
Shifflett, *Coal Towns: Life, Work and Culture in Company Towns in Southern
Appalachia, 1880–1960* (Knoxville: University of Tennessee Press, 1994);
Williamson Daily News, December 10, 1920; *1920 Annual Report of the Department
of Mines*, 12, 60.

41   *1920 Annual Report of the Department of Mines*, 12, 60.

42   *1920 Annual Report of the Department of Mines*, 19, 34–43.

43   *1920 Annual Report of the Department of Mines*, 59, 25, 19.

44   Allen W. Batteau, *The Invention of Appalachia* (Tucson: University of
Arizona Press, 1990), 125; *Mingo Republican*, August 23, 1917; James P. Johnson,
*The Politics of Soft Coal: The Bituminous Industry from World War I through the New
Deal* (Urbana: University of Illinois Press, 1979), 19, 45–51.

45   *Mingo Republican*, August 23, 1917; Johnson, 50.

46   Johnson, 50; Diner, 236.

47   *Williamson Daily News*, June 28, 1917.

48   *Mingo Republican*, August 23, 1917; Not all of Mingo's political leaders
sympathized with the operators' "plight." S. D. Stokes to J. A. Ferrell, September

19, 1917, Stokes Papers, WVRHC.

49 Johnson, 54; Despite Garfield's lack of qualifications, some in the industry admired him. One, George C. McIntosh, a West Virginia newspaperman and coal industry propagandist, eventually worked for Garfield. McIntosh Papers, WVRHC.

50 *Williamson Daily News*, August 14, 1920.

51 P. B. Noyes to Wells Goodykoontz, September 21, 1917; Copy of letter sent by Goodykoontz to S. D. Stokes, Stokes Papers, WVRHC; *Mingo Republican*, September 27, 1917.

52 *Mingo Republican*, October 11, 1917.

53 *Mingo Republican*, October 11, 1917.

54 *Mingo Republican*, November 8, 1917.

55 Johnson, 58, 64.

56 Johnson, 64–65; E. F. Striplin, *The Norfolk and Western: A History* (Roanoke, VA: Norfolk and Western Railway Company, 1981), 144–145. The inability of Mingo's "wagon mines" to capitalize on price increases because the N&W would not or could not provide a sufficient number of box cars is just one example of the limitations the railroad placed on the county's mines.

57 Johnson, 63.

58 Source for "equitable": Johnson, 63; Source for Barnes: *Progressive West Virginians*, compiled by John W. Kirk (Wheeling: Wheeling Intelligencer, 1923), 125.

59 *Mingo Republican*, January 7, 1918, and January 24, 1918.

60 White, Meade, and Juhling were all well-known "bolters" from Mingo's Republican ranks. See author's dissertation chapters three and six for details.

61 S. D. Stokes to the Honorable Lon Kelley, May 24, 1918, and Honorable Lon Kelly to S. D. Stokes, May 25, 1918, Stokes Papers, WVRHC.

62 See author's dissertation chapter six for a discussion of the Williamson Charter bill and the role of Stokes and York in Williamson's municipal politics.

63 Stokes-Kelly correspondence, Stokes Papers, WVRHC.

64 *McDowell Recorder*, May 21, 1918; See author's dissertation chapter six for the discussion of the War Eagle and Rockhouse "election fraud" cases; Kelley-Stokes correspondence, Stokes Papers, WVRHC.

65 Johnson, 71, 73.

66 Johnson, 71, 73.

67   Johnson, 71, 73.

68   Unsigned corporate letter to S. D. Stokes, July 9, 1918, Stokes Papers, WVRHC.

69   Johnson, 88, 91; McAlister Coleman, *Men and Coal* (NY: Farrar & Rinehart, 1943), 96; *Williamson Daily News*, February 1, 1919.

70   Coleman, 96; Johnson, 92.

71   *Williamson Daily News*, May 10, 1919.

72   Soule and Carosso, 513; *Williamson Daily News*, May 18, 1919.

73   *Williamson Daily News*, July 7, 1919.

74   *Williamson Daily News*, 2 July 1920 and 14 August 1920.

75   This does not mean that the operators objected to the imposition of martial law or the presence of federal troops in the strike district, because the restoration of order benefitted the operators' effort to resume a normal production schedule.

4

WORLD WAR I AND THE RISE

OF CLASS TENSIONS

"We hope our people can get along better in the future and
strive to have a better community in which to live."[1]
— *Williamson Daily News*

FOR MOST OF THE PERIOD between 1917 and 1919, World War I dominated
the social atmosphere of Mingo County. Civic organizations and public
activities focused almost completely on supporting the war effort.
However, government-sanctioned compulsory patriotism exacerbated
existing social strains, such as the county's ongoing struggles with
criminal behavior, sporadic acts of violence, and public health crises.
Although state and federal war policies facilitated the seizure of control of
Mingo's public sphere by the coal elite, the county's social transformation
proved temporary. The war not only empowered the "better classes," it
also inspired marginalized groups to seek greater autonomy. Ostensibly
united to fight the nation's enemies, Mingo Countians emerged from the
war with disparate and conflicting views of what the future should hold.
This new agitation, along with the cumulative effect of Mingo's systemic
political and economic dilemmas, facilitated the descent into violence in
the spring of 1920.

To most elites, America's entry into the war justified the adoption of
a centralized, paternalistic public policy that validated the repression of
"individualism and diversity of opinion in order to secure . . . unwavering
allegiance." West Virginia's application of restrictive war policies and programs was among the most extreme in the country. The implementation
of state and federal war regulations fulfilled the long-delayed dream of en-
forcing a rationalized work "discipline" in the Tug Valley. By cloaking their

efforts at social and economic control in patriotism, Mingo's political and industrial leaders legitimized the suppression of dissent and deviation.[2]

The wartime activities of Mingo's elite were celebrated in the *Williamson Daily News* and the formal reports of state officials. Mingo's Red Cross, YMCA, and Salvation Army campaigns were guided by political and economic leaders from Williamson. Para-political organizations such as Mingo's County and Community Councils of Defense and Four-Minute Men also featured bipartisan membership and prominent participation from coal company superintendents. Half of the 12-member County Council of Defense was linked to the coal industry.[3]

Although the war effort ultimately elicited an unprecedented bipartisanship from Mingo's political leaders, local disputes initially interfered with the establishment of the most significant war committees. The first crisis involved the county's Selective Service Board when turmoil threatened Mingo's ability to meet its quota of inductees. Other war committee work was also affected because the State Council of Defense appointed county and community Councils of Defense based on recommendations from the local conscription boards. When two overhauls of the committee failed to eradicate the intrusion of petty local squabbles, a leader from each party stepped forward to stabilize the situation. When anti-Hatfield Republican and coal attorney Harry Scherr of Williamson joined Williamson's deposed seven-term Democratic mayor A. C. Pinson on the draft board, matters settled down to the business of organizing Mingo's war effort.[4]

The domination of the Mingo draft board by Democrats and Williamson Republicans influenced the selection of the county's next important war committees. The state Council of Defense appointed the County and Community Councils of Defense based on recommendations from the county draft boards. As a result, Matewan Democrats and members of the Hatfield machine played no significant role in the war committee work of Mingo County. Despite being the second largest independent town in the county, Matewan had no representative on the County Council of Defense. Citizens of Williamson held eight of the twelve positions on the council. William N. Cummins, the superintendent of Red Jacket served on the County Council as the member from the Magnolia district. Democrat Dr. W. F. McCoy chaired the Matewan Community Council of Defense.[5]

The lack of Hatfields and Hatfield associates in Mingo's patriotic organizations coincided with their political eclipse from 1916 through 1920. As
corrupt as the Hatfield machine appeared to be, it had cultivated support
among the native working class and the ethnic communities of Mingo.
With the absence of the Hatfields and the corollary increased prominence
of coal men on the war committees, many people in Mingo County felt
unrepresented in those groups.[6]

Two weeks before America entered the war, the *Mingo Republican*
jested that the first American battle had already occurred in the streets
of Williamson. A group of local youths set upon and mercilessly beat
another young boy simply because his last name hinted at German heritage. Within weeks, however, the same paper somberly advised coal operators and their employees to be wary of transient miners. According to
the *Republican*, some of the "roving workers" were undoubtedly spies for
the enemy.

One Mingo County operator, Edward L. Sternberger, changed his name
to Stephenson out of a "patriotic" desire to support the war effort. The war
had legitimized the pursuit of public consensus and social conformity,
and anyone who did not accede to that demand was now suspect and subject to punishment.[7]

Great social pressure was exerted on the men of service age in Mingo
County. More than three hundred thousand West Virginians registered
for the draft. Of that number, 14 percent, or over forty-five thousand men
were mustered into service. Mingo County exceeded the state percentage
by inducting 15 percent of the residents who registered. On August 9, 1917,
the *Mingo Republican* published the names of the men selected in the first
draft under the headline "Honor Roll of Mingo County." One week later,
the *Republican* printed the "List of Ones Failing to Report." The August
23, 1917, issue contained the names of 36 "Alleged Slackers."[8]

Measures were taken to exert even greater control over the working
population in West Virginia in 1918. At the end of May, the *Williamson
Daily News* published an article whose very title reveals the parameters of
the renewed effort to harness people to the war effort. Mingo Countians
were told, "GO to War, GO to Work, or GO to Jail!" The article pinpointed
likely suspects for police roundups when it noted that Williamson's eight

pool rooms would be watched and that "jitney drivers" transporting "lewd women" after 9 p.m. would be arrested and prosecuted. Imposing limitations on acceptable leisure activities were followed by the extension of work surveillance. In July 1918, the state Council of Defense requested all employers in the state to submit weekly reports on employees who failed to work a full 36-hour week.[9]

Despite the elites' self-congratulatory public recognition, patriotic home front and war items periodically still shared news space with scandalous episodes of violence involving figures of authority. The three stories discussed here share thematic elements with the events of May 19, 1920, including violence to protect ones' manly honor, violence over politics, violence over a woman, or all of the above.

In September of 1917, a jury acquitted Matewan's former town sergeant John Hoskins of the 1916 murder of Henry Brewer. According to the *Mingo Republican*, the death of Brewer had resulted from a minor incident involving Hoskins, Brewer, and Brewer's brother-in-law, Lewis Hatfield. In May 1916, Hoskins stopped Brewer and Hatfield from driving over a water hose being used in a street-dampening, which Hoskins was supervising. As the three men exchanged words, Hoskins slapped Hatfield, whereupon Brewer hit Hoskins from behind. While Hoskins tried to regain his bearings, Brewer advanced on him with his hand in his hip pocket. Still reeling from the blow to his head, Hoskins shot and killed Brewer and then fled to Kentucky, where he remained for a year before returning to Mingo for trial.[10]

The recounting of the Hoskins-Brewer incident emphasizes several aspects of the approach to violence and justice in Mingo County. First, the actions of Brewer and Hatfield reveal that they were not intimidated by Hoskins' position as chief of police. Second, the escalation to physical violence occurred after Hoskins slapped Hatfield, which indicates that male honor became an issue in the exchange between the three men. Third, and most important, Hoskins' acquittal resulted from the jury's accepting his claim of self-defense; he faced no additional punishment as a result of his flight to Kentucky.[11]

After the acquittal, John Hoskins simply rejoined the community. Mysteriously, however, nearly two years to the day of Henry Brewer's death, men discovered Hoskins' lifeless body in Red Jacket's Mitchell Branch mine. Since no suspicious marks were found on the body, no one was charged in his death. Whether the Hoskins-Brewer incident led to Hoskins' death or influenced subsequent interaction between the families cannot be proven. However, it should be noted that during the strike period, 1920–1922, the Hoskins and Brewer families wound up on opposing sides of the conflict.[12]

Although Matewan and a few other communities had achieved notoriety for violence involving public officials, Williamson had escaped such infamy until the summer of 1918. At 9:00 in the evening on July 29, Jesse Huffman, the constable of Chattaroy precinct and an officer of the Sycamore Coal Company, shot and killed Williamson's Chief of Police John B. Maynard. Maynard's brothers, who were standing nearby, grabbed Huffman and arrested him. The *Williamson Daily News* concluded that Huffman's attack amounted to a crime of passion because "there seems to have been bitter feeling between the two men in which, as usual, women played their part, and may have caused the tragedy."[13]

Maynard's murder indeed may have resulted from a dispute over a woman, but a not-so-distant political controversy may also have contributed to Maynard's demise. Just one year before his death, Maynard had actively campaigned against the incumbent mayor, who was also a member of his party. At the time, Maynard allegedly was disgusted by "[the mayor's] long debauch, during which he tried and fined men for being intoxicated when he was worse intoxicated than the culprits." In return, the mayor, his brother-in-law Judge James Damron, and their political allies tried to fire Maynard. Maynard survived their machinations and remained chief of police even after the Democrats regained control of Williamson in 1917, only to be shot down a year later. Two months after the incident, a jury found Huffman not guilty of murdering Maynard, and all public discussion of the case abruptly ended.[14]

The last episode of violence involving public officials between 1917 and 1919 exposed the roots of the festering resentments that influenced the chain of events leading to the Matewan Massacre. According to the *Williamson Daily News*, on Sunday morning, December 7, 1919, Matewan's chief of police Sid Hatfield attacked Squire A. B. Hatfield. Ill-will between the two Hatfields stemmed from a controversy that had erupted between Mayor Testerman and Ance Hatfield, the manager of the Urias Hotel, earlier in the year. A. B. Hatfield had testified on behalf of a Urias Hotel employee who had successfully sued the police chief for improper arrest. The December incident between A. B. and Sid Hatfield spawned a flurry of rumors by "partisan friends," but the *Daily News* noted that despite the "lurid stories" circulating through the town, "it was apparent the Chief gave [the squire] a good pummeling." Although hauled into magistrate's court for feloniously assaulting A.B. Hatfield, Police Chief Hatfield escaped punishment because no one showed to press the issue.[15]

To the disgust of the *Daily News*, Chief Hatfield asserted that the incident had been misconstrued. Emboldened perhaps by the case's dismissal, Hatfield claimed that he had not attacked Squire, but had only interceded in an unfortunate barnyard incident. According to Chief Hatfield, Squire Hatfield had gone to feed his swine, and while bending over to slop them, "one of the pigs grabbed him by the nose and came very near biting it off," before he (Sid Hatfield) could rescue him. Hatfield's sly mockery inspired the *Daily News* to wearily editorialize in the article's conclusion that "we hope our people can get along better in the future and strive to have a better community in which to live."[16]

The violent exchange between Squire A. B. Hatfield and Matewan chief of police Sid Hatfield directly influenced the chain of events that led to the Matewan Massacre on May 19, 1920. First, despite sharing a last name with Sid Hatfield, Matewan's former mayor Andy Barrett Hatfield did not claim him as kin. As one Hatfield descendant observed, Sid probably was reminded of that rejection on a daily basis by the legitimate Hatfields of the town. Second, Sid had not been elevated to his position as Matewan's police chief by the Hatfields or their allies. By the time of Sid's appointment, a political nemesis of the Hatfield machine, Cabell Testerman, held Matewan's mayoral office. Third and last, when Albert Felts led a party of Baldwin-Felts agents into Matewan on May 19, 1920, he did so in direct violation of a peace bond issued by Sheriff Blankenship. The agents came with eviction warrants that Sid Hatfield later alleged were issued by A. B. Hatfield. One Massacre trial witness claimed to have seen Squire Hatfield watching the shootout with great joy.[17]

Two otherwise unrelated wartime developments illustrate the limits of elite control in Mingo County. As an unprecedented public health crisis, the Influenza Epidemic of 1918 exposed the superficiality of modern quality-of-life improvements in the county. Although programs designed to build consensus and conformity curtailed the social activities of Mingo's lower-class white population, the war facilitated unparalleled autonomy initiatives by two of Mingo's largest minority groups. The local communal memory of the flu epidemic underscores both the ability of Mingo Countians to come together during a crisis and the quiet heroism of the socially marginalized. By contrast, the role of the native white community in the erasure of the evidence of an independent ethnic presence speaks volumes about social hegemony in the county.

On September 26, 1918, the *Daily News* informed its readers that the "Spanish Influenza" had hit the nation's capitol and was poised to spread all over the country. Twelve days later, the newspaper reported that 480 people had already died at Camp Sherman, Ohio. Just two weeks after the *Daily News*'s first mention of the outbreak, the influenza "reached a violent stage in Williamson." In a chilling testament to the pitfalls of progress, the paper had chronicled the approach of the rapidly spreading disease. After appearing first in the state's eastern Panhandle, the flu traveled westward and southward along West Virginia's rail lines to Mingo.[18]

The flu's virulence and the speed of its diffusion forced the state's health department to shut down public intercourse—theaters, schools, and churches were closed, and public meetings and parades were strictly forbidden. Armed only with basic advice about illness and hygiene and alarmed by the ramifications of total public quarantine, the citizens of Mingo County held an open-air meeting on October 12 to discuss whether the community should comply with the state's proscriptions. At the time of the meeting's announcement, the *Williamson Daily News* reported that most of those afflicted were ill for three or four days and then rapidly recovered. Not until the epidemic had run its course did the disease's impact sink in.[19]

The state health department's report revealed the essential scope of the influenza's rampage through West Virginia. Based on data reported from every county, state health officials determined that the disease generally ran its course in seven weeks and that most of the fatalities stemmed from secondary illnesses such as pneumonia or meningitis. The flu lasted 43 days in Mingo, afflicting 1,158 and killing 144, almost six times the number of Mingo men who died while in service in the war.[20]

The flu disturbed the social and economic balance in Mingo. The most startling aspect of the spread of the disease was whom it seemed to target. Unlike other contagious diseases, which generally kill more of the very young and the very old, the flu killed those in the prime of their lives—from teenagers to men and women with young children. The mining community at Red Jacket felt especially exposed when the company doctor became the flu's first victim. After Dr. Goings' death, Mrs. Stella Presley observed that at Red Jacket, "about every other house had somebody dead in it." At the Glenalum mines, illness was so widespread that in order to keep the mines running, five young women, one of whom was a doctor's daughter, took up positions at the powerhouse, the switch, and the picking table, for "patriotic motives only." Modestly, the young women refused public acknowledgment of their sacrifice.[21]

The communities in Mingo were overwhelmed by the weather and the widespread sickness. Harry Berman noted that at Matewan, "ice was jammed up in the river fifteen, twenty feet," and yet people died so fast, Hawthorne Burgraff observed, "there couldn't be enough caskets brought in." Mrs. Eva Cook painted an evocative portrait of the atmosphere of death hanging over the county. According to Mrs. Cook, "the hearse was horse drawn with four lanterns . . . so many people died, they would bury them after dark . . . when you saw that hearse coming with those four lanterns lit . . . you knew it was an Influenza death."[22]

Stories recounted by survivors not only individualize and humanize the epidemic, they also illuminate elements of the community's social structure and how crisis challenged the bonds of family and community. Contemporary newspaper accounts reveal that a county-wide Influenza Relief Committee tried to organize assistance, calling in particular upon those with nursing experience and Boy Scouts. Survivors remember the efforts undertaken by individuals in the community. Clarence "Dutch" Hatfield remembered that "one or two old people . . . went house to house taking care of people." Daisy Nowlin stated that in the Matewan area, only "John and Mary Brown were not afraid and they would come . . . take

the families' washing . . . and see that they didn't want for anything."
Hawthorne Burgraff's aunt Elizabeth "Babe" Burgraff stands out as the
county's most colorful angel of mercy. Six feet tall and as "tough as a
man," (she once shot her own husband), Babe Burgraff cared for her extended family and claimed to have escaped the flu by drinking a shot of
moonshine "every once in awhile." The stories of these private citizens
who nursed others all celebrate the generosity of individuals who were in
other ways marginalized—the elderly, an African-American couple, and a
woman who did not obey the strictures of contemporary society.[23]

The rapid and tremendous devastation caused by the contagion was
not easily forgotten. Almost a year and a half later in the late winter of
1919–1920, a recurrence of the flu was reported near Mingo in the Tug
River coalfield. At about the same time, some of the operators in the
Williamson-Thacker coalfield raised company doctor fees. The manifest
insensitivity of this gesture magnified the growing distance between
Mingo's mining class and its employers. When asked to share examples
of the grievances that led to union agitation, one miner cited the ill-timed
hike in medical fees.[24]

One of the first miners to speak out in protest of postwar conditions
in Mingo County was W. E. Hutchinson, an African American. In fact,
Hutchinson was one of the two miners sent to Charleston to request
District 17's assistance in organizing Mingo County. For the last 20 years,
scholars have stressed the role of black miners in the struggle to unionize southern West Virginia. However, the full scope of the significance of
African-American involvement in the Williamson-Thacker strike has been
underappreciated. African-American miners were far fewer in number in
Mingo than in neighboring coal counties, and that it was they who rose
and demanded better treatment inspires a re-examination of the history
of race relations in Mingo County. What was uncovered indicates that, just
as in other African-American communities in the United States, World
War I unleashed long pent-up aspirations. The actions of Mingo's African-
American community during the war highlight the unintended corollary
effect of wartime propaganda. After the war, many of Mingo's African
Americans funneled their hopes into the union movement. The impact of
the 1920–1922 strike on Mingo's African-American community empha-

sizes the often overlooked local effect of such divisive conflicts.[25]

Table 12 provides statistics on the growth of Mingo County's African-American population. As the table shows, in 1900, only 309 African Americans lived in Mingo County, and constituted a mere 2 percent of the population. Although Mingo's black community expanded 400 percent by 1910, African Americans still comprised a mere 6 percent of the population of the county. Because so few African Americans lived in Mingo County, their low numbers exacerbated the significant challenges faced by blacks in West Virginia's singular sociopolitical system. Unlike their contemporaries in the South, African Americans in West Virginia retained the right to vote but faced a patchwork of inconsistent segregation practices. Several scholars have documented the experiences of African Americans who were lured to southern West Virginia's coalfields by the promise of higher wages and the hope of exercising rights denied them in other states. However, the experiences of African Americans in Mingo County demonstrate that opportunities for African Americans in West Virginia varied greatly, depending on the size of their community. In Mingo County, the low number of African Americans not only delayed community-building within the black community, it also complicated relations with whites.[26]

	1900	1910	1920
African Americans	309	1,236	2,191
% of total population	3%	6%	8%

Table 12. Population Growth of Mingo County's African-American Community, 1900–1920[27] Sources: "Abstract of the Thirteenth Census, with Supplement for West Virginia" and "Fourteenth Census of the United States, State Compendium for West Virginia."

Because almost all of Mingo's African Americans immigrated from elsewhere, there was no pre-existing community to welcome and succor the new arrivals. Moreover, they entered a community where even white migrants' access to power and influence depended on their ability to assimilate into the existing social hierarchy. Since their color automatically excluded them from intermarrying with the native population, pioneering

African Americans such as John Brown of Matewan were forced to construct other kinds of alliances with whites in order to safely establish residency. For John Brown this meant an alliance with the Republican leaders of the town, R. W. Buskirk and the allies of Greenway Hatfield. Brown's position and prosperity resulted from his own "wiles"—an ability to walk the fine line between accommodation and community leadership.[28]

Unfortunately, African-American loyalty to the Republican Party in Mingo did not mean as much as it did in other southern West Virginia coal counties. As we have seen, political power in Mingo County vacillated widely, not only between the Republicans and the Democrats, but also between factions of both parties. When the Democrats controlled the county, or even individual towns, the small flow of patronage to African Americans dried up. Moreover, because the African-American community in Mingo was smaller than in the neighboring coal counties, white Republicans in Mingo felt free to betray their most loyal supporters. African Americans in Williamson and Matewan were frequently caught up in the intrafactional disputes of their political patrons.[29]

The most significant obstacle to African-American community-building was the lack of a single center of African Americans in the county. Unlike John Brown of Matewan, most of the county's black population lived in company-controlled communities. An African-American neighborhood and business district eventually grew up in Williamson, but the lives of the "town" blacks were almost as circumscribed as those of the black miners.[30]

No evidence has been located to indicate that African Americans in Mingo County successfully launched community-building projects before 1917. World War I, however, precipitated a dramatic change when segregated war programs legitimized gatherings of black elites. African Americans from Mingo's towns and coal camps enthusiastically served in the county's Auxiliary Advisory Council of Defense and founded Red Cross chapters in Williamson and Red Jacket. Within a year, a non-war-related African-American organization appeared. In the spring of 1918, the Colored Boosters' Club held its first meeting in Williamson. Meetings consisted of debates and presentations by guest speakers such as State Commissioner of Agriculture J. H. Stewart. The coalescence of Mingo's

African-American leaders led to a dramatic and bold move in 1918. That year, community leaders, including John Brown of Matewan, renamed the Williamson Colored School the "DuBois School" in honor of W. E. B. DuBois. No records survive that would indicate whether the intent was to honor DuBois for his pioneering advocacy of racial pride and equality or for his leadership of African-American patriotic efforts during the war. However, even the obstinacy of the all-white Democratic county court could not intimidate the DuBois School trustees. When the county court refused to assist in transporting children to the school, Mr. Brown and the other leaders raised money to purchase a canvas-covered truck to ferry the students to school. By 1921, the name of the school had reverted to the Williamson Colored School in the West Virginia education directory. However, the self-empowerment symbolized by the brief life of the DuBois School bore fruit during the political and labor events of 1920, when African-American miners joined the strike and offered poignant explanations of its goals.[31]

In the spring of 1920, before the strike began, several African-American leaders started the Independent Colored Voters Club. Agitation for the miners' liberation soon coincided with this new unity-building activity among Mingo's African Americans. An indication of the connection between the two is the fact that one of the leading pro-union African-American miners, Frank Ingham, was married to the sister of the DuBois School's principal. However, the prominence of African Americans in the strike drew the attention of federal investigators; one casualty of the protracted bitterness of the strike seems to have been the united and independent activism of the African-American community. At a later meeting of the Independent Colored Voters Club, community leader and war veteran, Dr. James M. Whittico of McDowell County, was shouted down for expressing his continued support for the Hatfield machine and the Republican Party. African Americans John and Mary Brown and Maggie Washington were driven from Matewan by death threats from union supporters. Striking black miners also physically abused black replacement workers. The African-American community apparently never regained its wartime unity in the years after the strike.[32]

As in the case of Mingo's African-American community, World War I presented Mingo's second largest ethnic community with an opportunity to seek greater autonomy. In 1917 southern West Virginia's first cooperative mine was founded by and for Hungarian-American miners. Backed by Martin Himler, a Hungarian-American newspaper publisher and philanthropist who had worked briefly in the coal mines of Ohio, the Himler Coal Company operated for 11 years on the West Virginia-Kentucky border. Himler started the company with two objectives: to realize "the ideal of cooperation between labor and capital" and to "Americanize" Hungarian miners. The story of the Himler mine and the independent community it created reflects the impact of both the labor strife and the decline of the coal economy in the 1920s.[33]

As the prototypical eastern European immigrant group, Hungarian Americans have occupied a singular place in ethnic, labor, and coal history. Known for their determination to work under any conditions, Hungarians gained a reputation for being resistant to unionization and their willingness to work as strikebreakers. A primary reason for the Hungarian miners' disinterest in unionization allegedly stemmed from their disinclination to put down roots in America. A subsidiary goal of the Himler experiment centered on the desire to encourage Hungarians to invest in business ventures in America and not send all of their money back to Hungary. The Hungarians' work ethic would often draw unwanted attention during labor conflicts. The same traits that provoked animosity from co-workers elicited praise from their employers and other elites. "Old timers" in Mingo claimed that other miners hated the Hungarians because two on a section crew of eight "would work the other six to death."[34]

Exactly why Martin Himler and company chose a remote corner of Mingo County for their socioeconomic experiment is not clear. While Hungarian miners were first noted in the Big Sandy Valley in the 1890s, Table 13 illustrates that Mingo County attracted far fewer Hungarians than its neighboring counties. In Mingo County, Hungarians never exceeded 9 percent of the mine employee total, while in McDowell County, they comprised 25 percent of the total.[35]

Hungarian Population In	Mingo	Logan	McDowell
1910	309	239	1,816
1920	206	1,154	1,409

Table 13. Comparison of the Hungarian Population in Mingo, Logan, and McDowell Counties, 1910–1920 *Sources: "Abstract of the Thirteenth United States Census and Fourteenth United States Census Compendium."*

The opportunity that drew Hungarian miners to the Himler cooperative venture in Mingo centered on the promise of becoming worker-owners, since every employee was a stockholder in the company. In addition to this unique economic opportunity, the Himler Coal Company also constructed a model village. The miners could occupy a company house equipped with the latest amenities including electric lighting and electric or gas heat, or they could build their own home. A bakery specializing in Hungarian breads and cakes also served the community. Himler provided the company with a printing press in order to produce reading materials in Magyar. The company subsidized education, paying for three additional months of instruction and requiring teachers to hold college degrees. After attending the regular school, children studied the Hungarian language every day for two hours. The secretary of the company offered night classes in English and civics for the adults of the community. Himlerville even had its own bank.[36]

Despite what it offered its miner-owners, the Himler experiment had little chance of surviving. Constructed at the peak of wartime inflation, the Himler plant's modernity could not offset the disparity between production costs and post-war market prices. The company had budgeted a quarter of a million dollars to launch the operation, but ultimately spent three quarters of a million dollars. One of the largest expenditures incurred by the young company was the erection of the Warfield-Kermit bridge which took place between 1919 and 1921. Initially only one of several underwriters of the project, the Himler company eventually had to shoulder most of the construction costs. The other investors, also owners of coal lands or coal operators in Warfield and Kermit, were distracted by the strife that accompanied the Williamson-Thacker strike. Along with the unfortunate swings in the economy, the full-scale launch of the Himler mine coincided with one of West Virginia's worst labor struggles.[37]

By all indications, the Hungarian miner-owners of the Matta-May (Himler) cooperative mine did not join the strike. However, their indifference failed to protect the operation from the attendant violence. On November 10, 1920, someone blew up the drum and set fire to the headhouse of the Matta-May mine. Bloodhounds tracked a scent to the homes of Ira Maynard and Flem Stafford, who later denied participation in, or knowledge of, the assault. Although Maynard and Stafford were not linked to the UMWA in the public record, a minor detail of the story indicates that the men were union supporters. At the magistrate's hearing, attorney Thomas West, the local UMWA lawyer, represented Maynard and Stafford.[38]

A newspaper account of Himler's expansion in the spring of 1921 provides insight into the awkward position occupied by Hungarians in Mingo. The *Logan Banner* snidely dismissed Martin Himler's ambitious vision as "a little group of Hungarian coal miners" who had taken upon themselves "the task of healing the industrial wounds of West Virginia's bituminous coalfields through a cooperative plan." The pro-union miners engaged in Mingo's strike were equally hostile.[39]

Prior to the attack on the Matta-May plant, the union's key strategy in organizing Mingo depended on shutting down the mines. Miners who continued to work threatened that plan and were subjected to intimidation. One group that was targeted early, just one week after the Massacre, was the Hungarian miners employed by the Crystal Block Coal & Coke Company. Crystal Block's superintendent J. M. Tulley informed Governor Cornwell that the Hungarians were warned that they would be "riddled" with bullets if they persisted in their defiance of the strike. Although the Hungarian miners continued to work, they told Tulley that they were afraid to continue ignoring the union miners' threats. After the strike call took effect on July 1, 1920, coal production in Mingo stopped almost completely. By November, when the Himler plant was attacked, the mines in Mingo were almost back to full production, and frustration among the union miners and their sympathizers was high. In early December, a union miner was arrested for attacking John Florez, an employee of the Matta-May mine. According to the report on the incident, Will White beat and kicked Florez "mercilessly" because he was employed at a non-union mine and refused to quit working.[40]

Strike-related violence and intimidation failed to chase the Hungarian miners out of Mingo. After hitting a ten-year low of 127 miners in 1920, by 1923 their number had rebounded to 185. The Himler experiment also briefly survived the strike, but on May 22, 1925, the accumulated effect of cost overruns, an industry-wide recession, and the chaos of the strike years drove the company into receivership. Despite a valiant attempt to save it, four years later Himler's West Virginia assets were auctioned off, the new owners repaired the mines, and American miners moved in. Across the river in Kentucky, at the Hungarian village of Himlerville, the bank failed, the schoolhouse burned, and the houses were sold and torn down. The experiment described by Martin Himler as "the greatest single movement ever undertaken in this country" failed just ten years after its launch during the heady atmosphere of war.[41]

The key to understanding the violence visited upon Mingo's African-American and Hungarian communities by union activists during the Williamson-Thacker strike lies in examining World War I's effect on the miners of Mingo County. Exhorted and bullied into producing coal for the war effort, Williamson-Thacker miners emerged from the war convinced that they deserved a share of the prosperity created by their sacrifices. When their employers failed to deliver, the miners rose in protest and attacked anyone who appeared to oppose their goals.

In contrast to their employers, Mingo's miners had not materially benefited from the heightened demand for coal early in the war. The war boom had not resulted in better employment opportunity or security for Mingo's miners. The number of pick miners declined 91 percent before rebounding slightly in 1917, while in four years the number of machine miners increased by a meager 4 percent. By contrast, the number of men who worked as laborers inside the mines rose consistently. The increase in unskilled labor positions and the fluctuation in work opportunities for pick and machine miners undermines the idea that increased coal production caused by wartime demands translated into better employment possibilities for the average mine employee. In fact, for the men with mining skills, the immediate future seemed far from sure.[42]

The lack of stability in employment was compounded by the significantly increasing gap between food prices and income. The average miner's wages increased 7 percent between 1914 and 1917. Food prices rose an average of 30 percent between 1914 and 1917; the price of staple items like potatoes rose more than 50 percent. In sum, the cost of food outstripped the rise in miners' wages by nearly 400 percent. Compounding the gap between wage hikes and the price of food was the drop-off in opportunities to work—the average number of days during which mines in Mingo operated fell by 10 percent during the same period. Lack of certainty in job opportunities, fewer work days, and a higher cost of living all combined to lower the standard of living for Mingo's miners.[43]

The nation's entry into the war seemed to signal a turnaround in conditions, as the conscription of millions of American men dramatically affected labor relations. Around the country, and even in Mingo County, increased competition for workers necessitated a corollary improvement in wages and workplace and living conditions. The global scope of the war also inspired the funding of a wide variety of patriotic indoctrination or "Americanization" programs. In addition to undertaking these initiatives, however, and in collusion with West Virginia's governor, the industrial elite also imposed compulsory work restrictions that clearly defined the miners' place in the war effort. Although the operators intended to thus construct a pacified and biddable workforce, the uneven implementation of the improvements, combined with omnipresent coercive policies, only heightened the miners' desire for a more autonomous voice in determining the material betterment of their lives.

The drafting of able-bodied men into the army heightened competition for miners throughout West Virginia. The threat of losing their employees to unionized districts compelled non-union operators to offer inducements such as higher wages and better living conditions to their miners. In the Williamson-Thacker coalfield, this meant that the companies that had perpetuated the "rough and tumble" existence of the early years of mining were finally forced to institute improvements in and around their mines. Undertaken by the Williamson-Thacker operators as a group and as individuals, the policies' implementation underscores important trends in the coalfield's transformation during the war.[44]

First, and most important, miners' wages improved as a result of America entering the war. On May 1, 1917, wages in the Williamson-Thacker coalfield went up 10 percent, the third pay raise in just four weeks. In an attempt to thwart unionization, companies who could afford it awarded wage increases throughout the war period. In Mingo, wages rose only three more times between 1917 and 1919, but that translated to a 50 percent increase over the 1914 rate. By 1919, some miners in the Williamson-Thacker coalfield were earning close to seven dollars a day, which equaled the pay of union miners in the Kanawha coalfield.[45]

Both traditional and more recent studies of the company town system in southern West Virginia have documented that the war dramatically affected the appearance of, and living conditions in, the area's coal communities. In vying for skilled miners, even the most unenlightened operators were forced to improve the "villages of shacks" that had dotted the landscape since the pioneer days of mining in Mingo County. One such operator, Edward L. Stone of the Borderland Coal Company, reluctantly approved first the painting of his company houses, and ultimately, the construction of brick homes. The return on Stone's investment, according to his on-site manager, L. E. Armentrout, was the assurance that the quality of Borderland's housing exceeded that available in 90 percent of the Williamson-Thacker coalfield. When the New Howard Colliery opened in the winter of 1919, its miners were housed in dwellings that would have been recognizable to most contemporary Americans. Located on carefully laid out streets, the houses were plastered, and had electricity, porches, and washrooms.[46]

In conjunction with the efforts to improve their miners' physical environments during the war, operators of Mingo County also started programs to foster the development of community and company loyalty. Following the lead of the corporate owners of the "model" company towns in Logan and McDowell counties, several companies in Mingo sponsored an array of community beautification and educational enhancement projects between 1917 and 1919.

Analysis of the costs and benefits of annual lawn, vegetable, and flower garden contests reveals how the operators justified such non-production-oriented expenses. The companies often provided the seeds for grass,

vegetables, and flowers and awarded cash and other prizes to the winners of the competitions. In return for an outlay of a few hundred dollars,
the companies instituted an incentive-based form of social control. The
beautification of their camps helped attract a "better class of workers."
The operators believed, or at least hoped, that miners and their wives who
entered the contests would be too busy caring for their gardens and lawns
to drink and engage in rowdy behavior. Participation by foreign-born miners and their families also facilitated their "Americanization" by drawing
them into the local community of workers.[47]

On August 16, 1917, the *Mingo Republican* recounted the results of the
Crystal Block Coal & Coke Company's first garden contest. Fifty gardens
and lawns were entered in the contest. First place winners received $15
apiece and second place winners were given $10 each. The winning con-
testants were Steve Doby, Mike Liptic, Pat Browning, Andy Verba, and
Andy Bakos, four of whom were foreign-born. The *Republican* lauded the
officers of Crystal Block for sponsoring the competition and expressed
hope that other companies would emulate its example.[48]

In the interim, West Virginia's State Department of Agriculture drew inspiration from the garden competitions. In 1917, Mingo County Extension
Agent H. W. Prettyman served as one of the judges in the Crystal Block
competition and pledged to advise the miners for the 1918 competition.
However, the Department of Agriculture soon initiated its own garden program as part of a wartime initiative to promote self-sufficiency and food
production among the state's residents. The department's *Biennial Report*
revealed that across the state, extension agents worked with thirty-four coal
companies to encourage coal mining families to grow their own food. The
program summary also provides insight into the coal companies' contributions to the effort. According to the report, many of the companies fur-
nished the land, plowed and fenced it free of charge, provided fertilizer and
seeds at cost, and distributed prizes ranging from $60 to $150.[49]

The most aggressive and yet beneficent programs launched in Mingo
during World War I focused on education. Within six months of America's
entry into the war, the Howard Colliery underwrote an extensive renova-
tion of the Chattaroy school. The company paid for grading the school
lot, building a sidewalk, and installing sanitary drinking fountains and

a $600 playground apparatus. To ensure protection of the company's investment in the Chattaroy school, Howard Colliery also supplemented the principal's salary. According to the newspaper story on the Chattaroy school project, Howard Colliery received not only local approbation but statewide attention.[50]

The most ambitious education project undertaken by a coal company in Mingo County was launched in mid-November 1919. One of the Norfolk & Western captive mines, the Vulcan Colliery, announced its intention to offer night school courses for its employees. Although segregated, the night school program was open to both whites and African Americans. Far ahead of the rest of the Williamson-Thacker coalfield, the Vulcan night school program followed the most progressive efforts coming in to vogue throughout American industry.[51]

The story of the school and educational improvements in Mingo illuminates the many layers of the "benevolent paternalism" unleashed during World War I. On the surface, providing good schools helped the companies attract miners with families, believed to be more sober and reliable employees. In return, the companies not only gained a stable workforce, they also received "good press" for their concern and voluntary contribution to the local community. At the deepest level, subsidizing the salaries of teachers and principals also served a dual purpose. Dependent educators could be relied upon to promote values endorsed by the companies. Through the schools, the next generation of miners and miners' wives could be indoctrinated. Underwriting the cost of education eased the local tax burden, but more important, it could also be cited during the periodic outbreaks of complaints about the coal companies' comparatively light tax liability. By helping to provide good schools for community children, the coal companies thus also promoted an informal political subordination that benefited their interests on several levels.[52]

A minor scandal that erupted in Matewan illustrates the impact of the intertwining of local politics, education, and company paternalism. In April 1918, Professor N. L. Chancey, the superintendent of schools in the Magnolia district, suffered public disgrace when he admitted to "filching" funds from the Magnolia Board of Education. Part of the Hatfield machine and the closest associate of Matewan's recently defeated incumbent mayor A. B. Hatfield, Chancey might have escaped public censure with a gesture of restitution had he not roused the ire of several large coal corporations in the district.[53]

The *Williamson Daily News* revealed that despite Chancey's vow to replace what he had stolen, an investigation was pending because of the intervention of local coal companies. The *Daily News* story implied that the companies' action stemmed from an ongoing dispute between Chancey and the companies over the length of the district's school term. From the article's tone, the companies' officers believed that Chancey, while rejecting demands to expand the term, had pocketed the money which would have funded the extension.[54]

A single common factor unites all of the examples of the expansion of company paternalism in the Williamson-Thacker coalfield during World War I—in every case, the companies were among the largest in the field, and most were subsidiaries of the great corporations that mined vast deposits of coal in southern West Virginia. The projects sponsored by the Pond Creek Coal Company, Crystal Block Coal & Coke, Vulcan Coal Company, and Red Jacket in Mingo County conform with findings that assert that the most ambitious community-promoting projects were undertaken by the larger coal corporations.[55]

The end of the war did not mean the cessation of corporate experiments to "construct" communities. In 1919, the Solvay coal interests, which had bought the local mines of coal entrepreneur E. L. Bailey, started publishing a company magazine, *Solvay Folk*, ostensibly in the interest of, and for the "benefit" of its employees. The *Williamson Daily News* lauded the decision of Solvay's management to start the magazine because the company's size and ownership of mines in several coalfields made its leadership seem distant and impersonal. The *Daily News* asserted that *Solvay Folk* would "bring the workers of this big coal combination closer to the management." Solvay entrusted the magazine's content to a veteran West Virginia newspaperman, George C. McIntosh, who coincidentally, during Word War I, had also devised the Fuel Administration's "patriotic production" campaigns in which wounded veterans traversed the coalfields, exhorting miners to do their duty.[56]

The material improvement of the lives of Mingo's miners and their families during World War I, however, did not produce the anticipated results. Often, only those who worked for the largest and/or newest companies actually enjoyed the benefits of luxuries such as electricity and indoor plumbing. In the case of Borderland, the company merely built new and better houses for some of their workers near the older structures. New companies founded during the war boom often lacked the funds to maintain the communities in the postwar recession. The disparate living and work conditions that coexisted in the Williamson-Thacker coalfield contributed to the miners' discontent. Discrimination even within the most benevolent companies irritated. Only those who ranked above bank boss had special privileges at the store, and access to the amenities the companies allegedly took such pride in offering was restricted by the same rank. As one contemporary academic investigator noted, "It is painful to observe how little consideration the operators give their employees. They regard them so much as mere underlings who have no rights as members of a human brotherhood." When labor strife erupted in the Williamson-Thacker coalfield, the miners who most vehemently and violently rebelled against their employers' anti-union stance lived in, or in the vicinity of, the communities that had most aggressively pursued "paternalistic" policies.[57]

The subsequent rebellion of the miners who had received so much beneficence from their employers emphasizes the unfortunate timing of Mingo's adoption of benevolent paternalism, a practice that coincided with the imposition of state and federal mandated policies designed to guarantee steady production and worker compliance. The state government of West Virginia responded to the declaration of war by calling an extraordinary session of the legislature and enacting a body of laws that openly and deliberately marshaled West Virginia's citizenry for the war effort. Proud to have passed the nation's first compulsory work law, Governor Cornwell, the state legislature, and other civic and business leaders used the war crisis to promote a more rigid concept of social order and discipline than would have been acceptable in peacetime. Mingo County's application of two of the laws passed during the special session illuminates how West Virginia's coal communities in particular were harnessed to the war effort. An incident that occurred in a "wild" and "frightening" section of the county underscored that it was indeed the miners who were being controlled and that those who refused to conform to the new order were reviled as deviant remnants of an embarrassing past.[58]

First, in response to the nationalization of the state's National Guard units, the West Virginia legislature passed a law requiring each county to deputize 10 to 100 men to serve as public safety officers. Mingo County appointed 97 deputies, but just four of Mingo's seven districts garnered 75 percent of these special officers. The four districts—Lee, Magnolia, Williamson, and Stafford—were the political and economic centers of the county. Lee, Williamson, and Magnolia districts possessed over half of the county's coal mines and received two-thirds of the new deputies. Despite its reputation as a center for lawlessness and danger, Harvey District was assigned just eight deputies. Red Jacket and Borderland, the centers of production for the county's two largest coal companies, received nine deputies: five for Red Jacket and four for Borderland. Borderland, which had used the services of Baldwin-Felts detectives since 1915, now also received state and local subsidies for the surveillance and intimidation of its employees.[59]

On the morning of August 21, 1918, near Breeden in the Harvey district, a remote section of the county and home to half of Mingo's first 36 draft dodgers, two of the district's eight deputies surprised a "gang" of "army deserters," "slackers," and "sympathizers" who had come out of the mountains to eat at the house of a gang member's relative. While attempting to arrest two of the men, the deputies were shot and killed by the gang's leader, army deserter Albert McCloud. For nearly two years, Mingo's Sheriff G. T. Blankenship, U.S. marshals, and a unit of West Virginia "Home Guards" repeatedly tried and failed to subdue the gang.[60]

Throughout the months that Albert McCloud eluded the law, the citizens of Mingo were regaled with stories about the ways the criminals avoided justice. First, they drew on their familiarity with the terrain of that "wild and remote corner of . . . Mingo" to avoid capture. Second, the gang's families and other residents of the area provided aid. For example, parts of McCloud's abandoned uniform were discovered at Jack Marcum's farm—a place the sheriff and his deputies discovered had been turned into a bootlegger's "fortress" equipped with "subterranean passageways" between the buildings and gun "portholes" notched into the log walls.

Both Marcum and his wife were arrested for prohibition violations and shielding the deserters. In addition to arresting the Marcums, the deputies mocked them in the newspaper, telling the *News* that Marcum had protested the invasive search of a home in which he claimed to have raised 18 children and which looked to be a "sorry old hut" that measured only 12 by 14 feet, was about 50 years old, and looked as if "more than four-fifths of the family was still living there." The message was plain—McCloud and the Marcums were embarrassing holdovers from the past.[61]

When Albert McCloud surrendered on April 16, 1920, he defended his honor with passion, angrily denying allegations that he had been a moonshiner and had hidden for almost two years near his family's home. McCloud also indignantly declared that he had worked steadily at lumber camps in Lincoln County and had only briefly visited his family. McCloud's protests got little attention, and he soon disappeared from the newspapers. However, there are two noteworthy aspects to his story. The first is that for all the gung-ho patriotism of the coal and Williamson elite, they were unable to completely control the county's reaction to the war; at least some of the population resisted their lead. Second, given that it was well-advertised that the Harvey district was harboring draft dodgers and deserters, and that 17 lawmen had died there since 1900, it should have gone without saying that that section of Mingo needed more than just eight of the County's new special deputies. That the deputies were at the mines spoke volumes.[62]

The opportunity to truly capitalize on deputies "protecting" the mines and miners of West Virginia came when the state legislature passed the country's first compulsory work law on May 19, 1917. Officially titled the Idleness and Vagrancy Act, Senate Bill #7, popularly known as the "Slacker Law," criminalized unemployment. West Virginia's male residents, aged 16 to 60, were required to work 36 hours a week. Those afflicted with habitual intoxication and narcotic addition were not exempted, nor were professional gamblers or able-bodied men dependent on their wives and/ or children for support. State and local peace officers could challenge suspected violators of the law without waiting for sworn complaints. Private citizens were encouraged to watch for potential slackers; turning in loiterers was designated a patriotic act. Men convicted of "slacking" could be assessed a $100 fine and/or 60 days of public service work.[63]

Accounts of the law's enforcement and efficacy illustrate its effect in coal communities. Two of Mingo County's three municipalities responded to a survey about enforcement issued by the state Council of Defense. Matewan and Williamson reported one and twnty arrests respectively; in Matewan four slackers were forced to work, while Williamson sent seventy-five men into the labor force. Governor Cornwell's *Report on West Virginia in the War* revealed that a majority of state and local officers and industry leaders believed that coal production in the state increased because of the Slacker Law. According to the secretary of the state Council of Defense, many of the mayors who submitted compliance reports advocated the permanent retention of the law on the statute books.[64]

Although the leaders of the United Mine Workers of America had initially opposed America entering the war, the union eventually made great sacrifices to the war effort willingly. In return, the union was granted concessions that it would be reluctant to surrender in the war's aftermath. In what became known as the Washington Agreement, the nation's organized miners had agreed not to strike and thereby disrupt production for the war's duration. In return, they received a federally mandated wage increase. The UMWA also gained a small foothold in the New River coalfield when the Labor Bureau of the Fuel Administration ordered the coal companies in southern West Virginia to allow the miners to organize. The imposition of the Zone Plan during the war had also provided a brief respite in the ongoing battle between the unionized Central Competitive Field and the non-union Southern Appalachian coalfields. Federal regulation had temporarily halted the non-union assault on the markets traditionally served by the older organized coalfields.[65]

However, after the Armistice, UMWA members began to chafe under the continued restrictions of the Washington Agreement. Union miners had not only complied with its terms, they had willingly produced record amounts of coal. But when the Fuel Administration suspended regulation of the coal operators in February 1919, the government informed the UMWA that union miners were still prevented from striking until the end of the war or April 1920, whichever came first.[66]

Thus, after the Armistice, the UMWA and the men it represented felt betrayed. An inflationary cost of living had deflated their wages. The incentives offered in the non-union coalfields had angered many devoted union miners and eroded others' commitment to and confidence in their union. The dismantling of the Zone Plan also meant that competition for the Midwestern and Lake Cargo markets would probably resume between the non-union southern West Virginia coalfields and the Central Competitive Field. The final straw came in the summer of 1919, when the operators in the New River coalfield renounced their wartime agreement and set out to eliminate the union presence at their mines. By the time union delegates convened in Cleveland in September 1919, the men they represented were spoiling for a fight.[67]

In 1916, John J. Cornwell had wrested the governorship of West Virginia from Henry D. Hatfield's chosen successor Ira Robinson with the assistance of the unionized miners of the Kanawha coalfield. Angered by the limitations of the Hatfield Agreement, the insurgent officers of the UMWA's District 17 ignored the State Federation of Labor's advocacy of Robinson and supported Cornwell's candidacy. In return Cornwell promised to assist in organizing miners in the Fairmont coalfield of north-central West Virginia and to eradicate the mine guard system. Once in office, Cornwell maintained contact with District 17's President Frank Keeney and Secretary Fred Mooney but repeatedly betrayed the interests of West Virginia's organized miners. In particular, when Baldwin-Felts agents attacked UMWA organizers in Clarksburg, Cornwell wrote to Keeney and apologized but refused requests to order local operators to stop employing the agents. Cornwell also tolerated operator misuse of the Slacker Law.[68]

By the summer of 1919, just as the wartime advances in organization were being crushed in the non-union southern West Virginia coalfields, the lines of communication between the politically left-leaning leadership of District 17 and the governor dissolved. Within months, Cornwell became a vehement enemy of his former allies and turned increasingly toward the Local's conservative and accommodationist leaders who had originally been ousted by Keeney and Mooney in 1917.[69]

In late June, Mingo's operators decided to celebrate the apparent end of the post-Armistice coal recession by turning its sixth annual meeting into a two-day event. After a daylong business meeting, the Williamson

Operators' Association's closing banquet featured entertainment by co-median Luke McLuke and addresses by National Coal Association Vice President J. D. A. Morrow and West Virginia governor John J. Cornwell. Despite the festive and optimistic atmosphere, the subjects discussed by both Morrow and Cornwell hinted at the crises to come.[70]

After warning the association of the danger posed by Mexican crude oil, Vice President Morrow exhorted the men present to foster a greater spirit of cooperation among the producers of coal. Morrow claimed that the road to prosperity depended on friendliness and trust among coal op-erators. Hastily formed in response to wartime exigencies, the National Coal Association could not withstand the hostility between the unionized Central Competitive Field and the non-unionized coalfields of southern Appalachia. In spite of the efforts of members such as Vice President Morrow, the association virtually collapsed by 1921.[71]

Unlike Vice President Morrow, Governor Cornwell spoke directly about issues close to the heart of the Williamson operators. Cornwell began with a congratulatory recitation of West Virginia's war record and proceeded to what he considered the ongoing crisis in the body politic. According to Cornwell, social unrest menaced the state and the nation because of the high cost of living and changed economic conditions. Cornwell asserted that the radicalism spawned in such situations could not be thwarted by force or coercion but by treating the aspirations of the working class with respect. Speaking specifically to his audience, the governor urged the as-sociation to support fair legal treatment, wages, and living conditions for their mine employees, as well as educational opportunities for their chil-dren. Cornwell asserted that only by prioritizing these issues could they defeat the well-defined effort to lead the organized labor movement in West Virginia into a radical Socialist party.[72]

During the operators' convention and in the weeks that followed, the actions of the Williamson Operators' Association proved that Cornwell had found a receptive audience in Mingo County. His exhortation to di-minish the attractiveness of unionism by paying fair wages, improving living conditions, and offering better educational opportunities elicited a sense of self-congratulation among Mingo's operators. Cornwell's words offered the final validation they were seeking. After decades of fighting for survival, the Williamson-Thacker coal operators looked forward to a peaceful prosperity. In the business meeting of the convention, the operators approved a resolution that called for the use of a standardized proof-of-age form. Less than two months after the gathering, the operators' association of the Williamson-Thacker coalfield also approved an eight-hour work day for their employees. Commenting on the decision, the *Williamson Daily News* noted that the Mingo operators had chosen this course of action without regard for conditions in the surrounding coalfields.[73]

Freed from federal constraints, in the late summer of 1919, the operators of the Williamson-Thacker coalfield believed they had finally immunized their employees against the lure of unionism. Having granted semi-monthly pay in 1913, the operators capped a six-year effort to improve living and work conditions for their workers by granting the eight-hour day. However, far from the banks of the Tug there brewed a conflict that in less than a year would undo all of their work. Deluded by their own benevolence, the operators of the Williamson-Thacker coalfield ignored the storm warning issued by the events of the early fall of 1919.

In the late summer of 1919, District 17 President Frank Keeney made a fateful decision. Ignoring the pleas of his fellow officer Fred Mooney, Keeney responded to the anti-union rollback efforts of the New River operators by ordering union organizers into Logan's Guyan coalfield. Despite a warning issued personally by Logan's Sheriff Don Chafin, Keeney sent the men who had openly criticized his conciliatory relationship with Governor Cornwell into the heart of non-union southern West Virginia. Not surprisingly, Chafin and his army of deputies met the organizers' train and sent them back to Charleston under escort. War had been declared in southern West Virginia.[74]

As the UMWA's delegates traveled to Cleveland, Ohio, for the 1919 national convention, rumors began to circulate in the Kanawha coalfield that miners and their families in the neighboring Guyan coalfield were being abused and even killed by mine guards. Allegedly incited by the reports of murder and violence, the union miners of the Kanawha coalfield armed themselves, gathered just outside of Charleston (the state capitol), and announced they would march to Logan County (the center of the Guyan coalfield), and liberate and organize Logan's miners.[75]

Although District 17 officers were probably the instigators of the march or at least co-conspirators in spreading the rumors, they went to great lengths to convince observers that the miners' gathering had resulted from a spontaneous eruption of mass indignation. Late on September 4, 1919, District President Frank Keeney and Governor Cornwell traveled together to the miners' gathering to urge the men to disband. Believing that his promise to investigate the rumors had convinced the miners to go home, Cornwell returned to Charleston. When the march started on September 5, Cornwell claimed that after he had spoken and left, radicals and socialists incited the men again. He also felt personally betrayed by Keeney, whom he felt had not done enough to stop the march.[76]

As the miners approached Logan on September 6, Cornwell telegraphed Newton Baker, the secretary of defense, and General Leonard Wood, commander of the Central Division of the United States Army, to warn them that the situation in the coalfields of southern West Virginia threatened to deteriorate beyond state control. Shortly after Cornwell informed Keeney that the army would be available for use in southern West Virginia, the armed march on Logan ended. The *Williamson Daily News* reported that on September 7, the miners gave up their "pilgrimage," disbanded, and returned home. On September 9, Cornwell wrote to Keeney to inform him that he had appointed Colonel Thomas B. Davis, West Virginia's acting adjutant general, to head a commission to investigate not only the rumored abuses in Logan County but also who was responsible for the assemblage of armed miners.[77]

As the commission prepared to launch its investigation, the national UMWA organization weighed in with its opinion of the situation in West Virginia. The delegates at the Cleveland convention authorized acting president John L. Lewis to send Governor Cornwell a telegram. On September 11, Lewis wrote to Cornwell:

If the Governor would exercise the authority and influence of his office to have the laws of West Virginia guarantee free speech and free assemblage and the right to organize it would not be necessary for freeborn American citizens to arm themselves to protect their Constitutional rights.

Cornwell did not respond to Lewis' telegram directly. However, as subsequent events during his administration illustrated, while publicly supporting workers' organizations, Cornwell actively opposed any actions undertaken by the UMWA in West Virginia.[78]

The resolutions passed during the UMWA convention of 1919 and the governor's commission investigation of the armed march on Logan profoundly affected the subsequent interaction of the coal companies and miners in southern West Virginia, and in particular Mingo County. Although the delegates at the UMWA convention voted to ban syndicalists from membership in the union, they also passed resolutions viewed as radical and dangerous by most coal industry executives. The policy committee endorsed nationalizing the mines, and the entire delegation supported a plan for nationalizing the railroads.[79]

The most decisive action taken by the UMWA convention was to officially challenge the federal government's assertion that the Washington Agreement remained in effect. The convention also issued a list of demands that sent shock waves through the coal industry and the federal government. The miners called for a 60 percent increase in wages, a six-hour work day, and a five-day work week, an end to the penalty fines imposed by the Washington Agreement, and a nationwide contract without sectional settlements. If the coal industry failed to grant these concessions, a national strike of coal miners would begin on November 1, 1919.[80]

Efforts to stave off the national strike ended miserably. Brought together by federal mediators, union officials and operators' representatives found no middle ground and abandoned negotiations. In West Virginia, although the commission impaneled to investigate the armed march issued a report of its findings, the actual investigation had not been completed. When the commission denied District 17's request for immunity for testifying miners, too few witnesses cooperated. Convinced that the commissioners sought only to blame the miners and not address the injustices occurring in Logan County, District 17 published a request for a federal investigation.[81]

A power vacuum in both the national government and the UMWA further complicated the descent into crisis in the fall of 1919. President Wilson suffered two strokes in October 1919, and his incapacitation allowed the hardliners in the administration to move decisively against the UMWA. By this time, illness had also forced UMWA President Frank J. Hayes to entrust control of the union to Vice President John L. Lewis. Lewis had moved rapidly up through the union ranks, and the events of 1919 allowed him to grasp even greater power. As one historian noted, Lewis "commanded the statistical data of the coal industry" better than anyone else. This knowledge and his own intractable nature made Lewis as determined as his opponents in the growing crisis. As the November 1 strike deadline approached, the men who were in a position to affect a compromise refused to budge.[82]

The state-level confrontation between District 17 President Keeney and Governor Cornwell reflected the belligerence of the national standoff. When Keeney informed Cornwell that West Virginia's union miners would obey the strike call, Cornwell responded by calling troops into the coalfields two days before the strike was supposed to begin. Congressman Wells Goodykoontz of Williamson declared that in striking, the union "paid no more attention to . . . [their contracts] than did the Germans to their treaty obligations with the Belgians." Goodykoontz implied, of course, that if the miners obeyed Lewis and Keeney, they were unpatriotic and disloyal.[83]

Once the strike began, Wilson Administration hardliners led by Attorney General A. Mitchell Palmer moved to destroy it. Palmer and his allies engineered a restoration of the Lever Act, which stipulated that any conspiracy to limit the supply or distribution of coal was illegal. Based on this clause and the terms of the Washington Agreement, Palmer got an injunction against the strike on November 8, 1919. After two days of deliberation, Lewis and the other UMWA leaders decided to end the strike, but the miners stayed out. Slightly less than a month after the injunction went into effect, Lewis and 83 other union leaders were arrested on contempt charges. At a White House conference on December 7, John L. Lewis agreed to call off the work stoppage. In return, the federal government agreed to establish a commission to study conditions in the industry and ordered an immediate 14 percent increase in miners' wages.[84]

Although the *Williamson Daily News* reported on the strike, no strike-related activity occurred in Mingo County. There were at least two possible reasons for the calm. First, during his testimony in the 1921 senate investigation, Fred Mooney admitted that District 17 officers rejected a request for organization from Mingo County in 1919. Second, during the strike, the miners in Mingo were making good wages with a greater opportunity to work. When the strike stalled work in West Virginia's union fields, the operators in the non-union fields seized the opportunity to increase their own output. Spurned by the union and offered the chance to make more money by their employers, the miners in Mingo worked on through the strike. In fact, at the strike's height, Mingo Sheriff G. T. Blankenship reported to Cornwell that since the mines were working at full capacity, he doubted there would be labor unrest in Mingo County "unless the operators [got] excited."[85]

The coal operators of the Williamson-Thacker coalfield and the conservative businessmen of Mingo County acted vigorously during the 1919 strike, and their response hinted at the issues that would be raised the following year. They wrote to Governor Cornwell to endorse his efforts against radicalism in the state. Several of the paternalistic programs discussed earlier in this chapter were started during this period of the general strike. Like Cornwell, the coal men of Mingo County claimed to support unionism in spirit, but when confronted directly by unionists, they denounced their opponents as radicals who were being swayed by "anarchistic teachings." Having followed Cornwell's advice regarding higher wages and improved living conditions, the operators of the Williamson-Thacker coalfield believed their employees were immune to the siren call of organization.[86]

Wells Goodykoontz's November 1, 1919, speech on the floor of the House of Representatives reveals the depth of the coal operators' self-delusion. According to Goodykoontz, only one out of ten West Virginia miners supported the 1919 strike because they happened "to have a favored lot, mining thick seams . . . under genial surroundings, at wages, in some cases, higher than the union scale." The UMWA's $15,000,000 campaign fund "to corrupt non-union miners" would not sway the miners of West Virginia because they were "the most happy . . . independent . . . and contented . . . of all our citizenry."[87] Seven months after Goodykoontz painted this rosy picture of life in southern West Virginia's coalfields, the miners in Mingo County defied their employers and flocked to join the UMWA.[88]

Statistics reported in the 1919 *Annual Report of the Department of Mines* offer the first clues to why the miners of the Williamson-Thacker coalfield would move to organize almost six months after the 1919 national strike. Between 1918 and 1919, the number of working mines decreased in Mingo, and although more men were employed, 8 percent fewer of the jobs were held by miners. The men worked nearly a month less in 1919 than they had in 1918, and although the average per-ton wage paid to the miners rose almost 30 percent by 1919, the operators' gross profits approached 600 percent. As illustrated in Table 14, in the period when operators' profits rose dramatically, miners' wages as a percentage of production cost fell. To the miners this meant that at a time when the cost of living due to wartime inflation was at all time high, their employers could have, but chose not to, pay them more.

Year	Average Wage*	Average Selling Price+	Wages as a % of Selling Price
1917	.59	1.84	32%
1918	.75	2.77	27%
1919	.82	2.56	32%

Table 14. Comparison of Average Wage and Coal Prices in Mingo County, 1917–1919 *Source: "Annual Report of the Department of Mines." *Average based on wages paid by car and ton. +Selling Price based on price per ton.*

These record profits resulted from the high selling price of coal in 1918. One need only compare prewar and 1918 dividends to appreciate the impact of the price inflation. For example, before World War I, the Borderland Coal Company paid average yearly dividends between 15 and 30 percent. In 1918, despite the cost of building new company houses and a tipple, Borderland paid dividends of 60 percent, the highest in the company's entire history.[89]

Despite the inflationary benefits to the companies and their investors in 1919, the Williamson-Thacker coalfield fell further behind its neighboring coalfields, the Guyan and Pocahontas coalfields of Logan and McDowell counties. A smaller percentage of miners lost ground to laborers in those coalfields. Like the Williamson-Thacker coalfield, the Pocahontas coalfield lost work days and mines, and also produced less coal than it had in 1917 and 1918. But to understand the different impact of those losses in Mingo and McDowell Counties, one has only to look at the scale of the losses in Mingo and McDowell. McDowell's loss of tonnage output in 1919 *exceeded* Mingo's output for the entire year.[90]

In contrast to Mingo and McDowell, Logan County added 16 new mines in 1919, miners worked nearly two weeks more, and they produced nearly a half million tons more coal. Unlike Mingo and McDowell, Logan benefited from the high wartime prices and special market conditions in the immediate post World War I period, which stimulated expansion of production in existing mines and encouraged the opening of many new mines.[91]

Comparison of the statistics from Mingo, Logan, and McDowell counties illustrates that the coal industry in Mingo County emerged from the war period less secure than Congressman Goodykoontz's speech led his contemporaries to believe. Although the war boom had brought improved wages and living conditions for many of Mingo's miners, these advances were undermined by a higher cost of living and by the perception that their employers were prospering even more than the miners themselves.

World War I precipitated the political, economic, and social cleavage that brought Mingo County into the cross-hairs of the conflict over the unionization of southern West Virginia. Empowered by their control of Mingo's wartime patriotic effort, the coal and Williamson elite believed they had finally vanquished the Hatfield machine and all it personified. In the name of patriotism, the elites imposed new forms of social delineation and control that exacerbated old class conflicts. The war also unleashed the ambitions of the county's minority population. By 1920, undercurrents of seething political, economic, and social tensions gathered, needing only the right spark to explode.

Mingo County had never been a socially homogenous community. Industrialization had not automatically imposed a modern bi-modal system with distinct class barriers. People in the county did not inhabit only isolated, locked-down, operator-dominated coal camps. In fact, just the opposite had occurred in Mingo County. Independent towns had grown up around the mines and provided an outlet for activities that defied corporate constraints. The kinship connections between miners and native elites also provided opportunities for some miners to participate in the public sphere in ways that have gone unacknowledged for decades.[92]

However, World War I set in motion a social transformation that crystallized during the 1920–1922 strike. First, administration of local wartime programs was dominated by the industrial elite and their Williamson allies. The correlative eclipse of the Hatfield machine fostered a sense among this group that control of Mingo County had finally come into the hands of the "better classes." At the same time, the war had also liberated the expectations of the county's working class and ethnic populations. In the spring of 1920 when UMWA representatives appeared in Mingo County, miners and African Americans responded enthusiastically to their dual message of unionism and political freedom.[93]

The degree to which the war divided Mingo County is illustrated by an event that occurred in the midst of the drama surrounding Logan and Kanawha counties during the 1919 coal strike. In late October 1919, Mingo County veterans launched the Eph Boggs Post of the American Legion. Members asserted that their goal was to enlist all of Mingo's veterans in the post, but the roster of the officers selected by the 35 men present at the first meeting reveals who claimed Mingo's patriotic war legacy. Mingo Keadle, John Chafin, and Hyman Crigger were the sons of a former Republican newspaper editor, Mingo's first county clerk, and a coal company official, respectively. Most important, the veterans selected Congressional Medal of Honor winner and Paint Creek strike terror A. A. "Bad Tony" Gaujot to be post commander. Two years later when anti-union community members started the Mingo Militia to assist the strike-busting efforts of the state police, six of the nine founding members of Mingo's American Legion post joined the anti-union vigilance committee.[94]

By the end of 1919, the local and external causes of the 1920 conflagration were falling into place; these factors combined to create the national, state, and local prelude to the events of May 19, 1920. In late 1919 and early 1920, as state and national political and economic conflicts coalesced, the coal elite and their allies possessed a tenuous hold over Mingo County. But all was not well. Wartime regulation had benefited the large corporations in the coalfields of Mingo's powerful neighbors, not the Mingo operators. Within Mingo, on the political scene, the Hatfields were gearing up for a hard-fought return to power, while the City Ring Democrats were trying to decide what to do with the "County" Democrats, whose lame duck nominal leader still served as sheriff. Tensions were rising, as was a belief in several corners that significant change could be achieved if only something could happen. The conflagration needed only the right spark.

NOTES

1   "Hatfield Vindicated," *Williamson Daily News*, December 13,1919.

2   Source for "diversity": Paul L. Murphy, *World War I and the Origins of Civil Liberties in the United States* (New York: Norton, 1979), 26–27; John Hennen, *The Americanization of West Virginia: Creating a Modern Industrial State, 1916–1925* (Lexington: University Press of Kentucky, 1996), 15; Gordon B. McKinney, "Industrialization and Violence in Appalachia in the 1890s," *An Appalachian Symposium: Essays Written in Honor of Cratis Williams,* edited by J. W. Williamson (Boone, NC: Appalachian State University Press, 1977), 131–144, 137.

3   John J. Cornwell and others, "West Virginia in the War," *West Virginia Legislative Handbook and Manual and Official Register for the Year 1918,* compiled and edited by John T. Harris, clerk of the Senate, (Charleston, WV: Tribune Printing Company, 1918): 931, 941, 944; The "coal" people on the county council were: George Bausewine, Jr. (officer in the Williamson Coal Operators' Association), William N. Cummins (superintendent of Red Jacket Consolidated Coal & Coke Company), and Mrs. Mark Russell (Mark Russell was a coal dealer in Williamson). These three were joined by pro-coal business leaders and investors E. F. Randolph, B. Randolph Bias, and Harry Scherr of the Coal City Club.

4   *Mingo Republican*, September 30, or August 30 [probable date-printing error] 1917; Source for reason for disruption: *Martinsburg Herald*, September 7, 1917; "West Virginia in the War," 793; For more on Pinson, see author's dissertation chapter six; For more on Scherr, see author's dissertation chapter three.

5   "West Virginia in the War," 793, 806.

6   This refers only to home front, non-military organizations. Several Hatfields, including former West Virginia governor Henry D. Hatfield, served in the military during the war. County court clerk and Hatfield Republican Elihu Boggs was by profession a coal miner. Fred Burgraff, coal miner, Matewan Massacre defendant, and union activist from 1920–1922, had served as a deputy sheriff under Greenway Hatfield from 1912–1916. See author's dissertation chapter six for details.

7   *Mingo Republican*, March 22, 1917; May 3, 1917; "Obituary of Edward L. Stephenson," *Williamson Daily News*, September 9, 1919.

8   "West Virginia in the War." By comparison, McDowell sent 16 percent and Logan sent 12 percent. Mingo's war casualties totaled 25: 10 were killed in action, 2 died from wounds, 4 died from disease in Europe, and 9 died in the United States from disease and other causes. Compiled from data in Jim Comstock, ed., "Casualties in World War I," in *The Soldiery of West Virginia, West Virginia Heritage Encyclopedia*, vol. 9, supplemental series (Richwood, WV: Jim Comstock, 1974), 229–272; *Mingo Republican*, August 9, 1917; *Mingo Republican*, August 16, 1917; *Mingo Republican*, August 23, 1917.

9   *Williamson Daily News*, May 30, 1918; "West Virginia in the War," 786.

10   *Mingo Republican*, September 13, 1917.

11   For more on self-defense arguments, see author's dissertation chapter five; For broader cultural application see: Montell, *Killings*, 149; also Roy E. Nisbett and Dov Cohen, *Culture of Honor: The Psychology of Violence in the South* (Boulder and Oxford: Westview Press, 1996), xvi, 26, 86.

12   *Williamson Daily News*, May 7, 1919; For more on the connections between the Hoskins, Brewer, and Chambers' families, see author's dissertation chapter 10.

13   *Williamson Daily News*, July 30, 1918.

14   Source for "debauch": *Mingo Republican*, May 31, 1917; *Williamson Daily News*, September 17, 1918.

15   *Williamson Daily News*, February 20, 1919; Shortly after taking office, Testerman had revoked the hotel's license on charges related to alleged illegal and lewd activities. Details of this incident appear in "Sid Hatfield, by Isaac Brewer," Lewis Collection, ERCA. Although Brewer errs in the chronology of some of the events, his story otherwise matches accounts from the *Williamson Daily News*, December 13, 1919. Source for "pummeling": *Williamson Daily News*, December 13, 1919.

16   *Williamson Daily News*, December 13, 1919.

17   Source for "kin": Hatfield correspondence, letters no. 17 and 29; Source for "nemesis": *Williamson Daily News*, July 7, 1919. Since Sid Hatfield had become Matewan's chief of police by the summer of 1919 it can be surmised, but not proven that he was appointed by Mayor Testerman; Source for "violation": *Williamson Daily News*, April 29, 1920. Albert Felts had been arrested on April 28, 1920, and placed under a $2,000 peace bond. Exactly who issued the eviction

warrants became a major dispute in the Massacre's aftermath. "Testimony of R. M. Stafford," March 15 1921, unknown newspaper, Matewan Omnibus Collection, ERCA; Source for "joy": Hugh Combs Trial Testimony.

18   *Williamson Daily News,* September 26, 1918; *Williamson Daily News,* October 8, 1918; *Williamson Daily News,* October 10, 1918; *Biennial Report of the State Department of Health, For the Years 1919–1920* (Charleston, WV: Tribune Printing Company, 1920), 12.

19   *Biennial Report of the State Department of Health, For the Years 1919–1920; Williamson Daily News,* October 10, 1918; *Williamson Daily News,* October 12, 1918.

20   *1920 State Health Report,* 12; and *Williamson Daily News,* October 10–November 22, 1918. Mingo County ranked nineteenth out of the state's fifty-five counties in the number of flu-related fatalities.

21   Stella Presley interview; *Williamson Daily News,* November 2, 1918.

22   Harry Berman interview with John Hennen, 1989 Matewan Oral History Project; Hawthorne Burgraff interview with John Hennen, 1989 Matewan Oral History Project; Eva Cook interview. Mrs. Cook was born during the epidemic; her mother fell ill while recuperating from the birth, but survived.

23   Several of the Matewan Oral History Project narrators discussed the fatalities that resulted from the epidemic: Mrs. Mattie (McCoy) Allara; Mrs. Bertha Staten; Mrs. Eva Cook; Mrs. Edith Boothe; and Mrs. Vicie (Hatfield) Simpkins Blackburn; *Williamson Daily News,* October 22, 1918; Clarence "Dutch" Hatfield interview with Rebecca J. Bailey, 1989 Matewan Oral History Project; Daisy Nowlin interview with Rebecca J. Bailey, 1989 Matewan Oral History Project; Hawthorne Burgraff interview.

24   *Coal Trade Journal* 51 (February 18, 1920): 174; "Testimony of W. E. Hutchinson," *West Virginia Coal Fields,* 79.

25   John Hope Franklin and August Meier, eds., *Black Leaders of the Twentieth Century* (Urbana: University of Illinois Press, 1982), 113–114. For more on this regional effort and its effect on race relations during the second mine war see: Joe William Trotter, "Black Miners in West Virginia: Class and Community Responses to Workplace Discrimination, 1920–1930," in *The United Mine Workers of America: A Model of Industrial Solidarity?* edited by John H.M. Laslett (University Park, PA: The Pennsylvania State University Press, 1996): 269–296, 294.

26   *Abstract of the Thirteenth United States Census, with Supplement for West*

Virginia, 592; Corbin, *Life, Work, and Rebellion* (1981); Ronald Lewis, *Black Coal Miners in America: Race, Class, and Community Conflict, 1780–1980* (Lexington, KY: University of Kentucky Press 1987); Trotter, *Coal, Class, and Color: Blacks in Southern West Virginia, 1915–1932* (Urbana: University of Illinois Press, 1990). The paradigm of African-American life in southern West Virginia was constructed based on the experiences of African Americans in Kanawha and McDowell counties. In McDowell County in the period under study here, African Americans represented between 25 and 30 percent of the total population, versus a maximum of 8 percent in Mingo. The author questions whether critical sociocultural experiences would have been that similar between such differently sized populations.

27   Readers are reminded that 1900 was the first United States Census for Mingo County. Sources: *Abstract of the Thirteenth United States Census, with Supplement for West Virginia*, 592; and *Fourteenth United States Census, State Compendium for West Virginia*, 27.

28   Corbin, *Life, Work, and Rebellion*, 62; Johnny Fullen interview and Harold Dickens interview with Rebecca J. Bailey, Summer 1990 Matewan Oral History Project. Several generations of Matewan residents, black and white, respected John Brown. When Mr. Dickens, a white former resident of Matewan, spoke of Brown using his "wiles," he was lauding Brown's ability to survive and thrive in an era of challenging race relations.

29   The assertion about race and politics is based on F. H. Evans' support for Jim Crow bills in the 1907 legislature and J. M. Studebaker's tacit endorsement of the firing of Williamson's African-American employees in 1917. For more on Evans, see: *Charleston Advocate*, February 14, 1907; For more on Studebaker, see *Mingo Republican*, May 31, 1917. The two best documented cases of this phenomenon are the Studebaker incident mentioned in the previous note and the 1910 Matewan election incident. In 1910, native elite-machine politician Greenway Hatfield, who with great cunning cultivated black political support, slapped a coal company doctor and fellow Republican for criticizing the way Hatfield was allowing the "darkies" to vote.

30   Cubby, "Transformation," 161; *Abstract of the Thirteenth Census, with Supplement for West Virginia; and Fourteenth Census, West Virginia State Compendium*. The aggregation of population statistics shows that the majority of Mingo's African-American residents lived in Lee/Williamson, Magnolia, and Stafford districts—the centers of commercial coal mining in the county.

Source for Vinson Street as the center of the African-American neighborhood in Williamson: Huey Perry, *They'll Cut Off Your Project: A Mingo Count Chronicle* (New York: Praeger Press, 1972), 54; Source for Third Avenue as the center of the black business district: Archie Bland interview with Rebecca J. Bailey, Summer 1990 Matewan Oral History Project.

31 Hennen, *Americanization*, 51; "West Virginia in the War," 815; *Williamson Daily News*, February 19, 1918; *Williamson Daily News*, April 4, 1918. Members included E. D. Britton, P. L. Hines, E. S. Campbell, and R. B. Hill; M.P. Shawkey, *West Virginia Education Directory, for the School Year 1917–1918* (Charleston, WV: Tribune Printing Company, 1918); and from an interview with Johnny Fullen, grandson of John Brown, of Matewan. The African Americans of Mingo County who would have been privy to this decision had all passed by the time the author learned of the DuBois School. Contemporary local newspapers and the West Virginia State Directory of schools were silent on the matter. Also, archivists of the W. E. B. Dubois papers were unable to locate any materials that might have revealed that DuBois had learned of this action; Johnny Fullen interview.

32 *Williamson Daily News*, May 15, 1920; "Testimony of Frank Ingham," *West Virginia Coal Fields*, 33; "Negro Activities," October 9, 1920, "Reports by Informant C-61 to A. E. Hayes, for the Southern District of West Virginia, October 9–October 30, 1920," in *Federal Surveillance of Afro-Americans (1917–1925): The First World War, the Red Scare, and the Garvey Movement*, edited by Theodore Kornweibel, Jr. (microfilm project of University Publications of America); *Williamson Daily News*, May 15, 1920; Source for John and Mary Brown: Johnny Fullen interview; Source for Maggie Washington: unknown newspaper, Lewis Collection, ERCA. Maggie Washington was a domestic servant in the home of Matewan Massacre defendant Art Williams; "Memoranda of Mingo Conditions Compiled Monday," undated, Cornwell Papers, WVRHC; "Dr. James M. Whittico," *Williamson Area Heritage Book*, 105; For more on John Brown, see: Johnny Fullen interview; For more on James Curry, see: James Curry interview with Rebecca J. Bailey, Summer 1990 Matewan Oral History Project; One indication that Mingo's black elite retained their commitment to racial issues came with the founding of Mingo's NAACP chapter in 1923, five years after Bluefield, but four years before Logan. Trotter, *Coal, Class, and Color*, 246.

33 Mary Lucille Chapman, "The Influence of Coal in the Big Sandy Valley" (Ph.D. diss., University of Kentucky, 1945), 224–226, 236. Chapman's source was

the *Paintsville Herald*, 24 January 1924.

34   Bailey, "Judicious Mixture," 150; *Logan Banner*, May 21, 1921; Source for observation about Hungarian miners: Hatfield correspondence, letter no.22.

35   Carol Crowe-Carraco, *The Big Sandy* (Lexington, KY: University Press of Kentucky, 1979), 91; Source for Mingo: data extrapolated from the *Annual Reports of the West Virginia Department of Mines, 1907–1921*; Source for McDowell: Bailey, "Judicious Mixture," 152. In fact, for much of the period under examination in this study, the Hungarian percentage of the mining population of Mingo hovered around 6 percent, one-fourth of McDowell's percentage; The attack took place in 1908—see author's dissertation chapter five for details.

36   Chapman, 230–233; In *The Big Sandy*, Carol Crowe-Carraco states that the Himlerville cottages contained "five rooms, two windows per room, tubs and showers and electric lights." Crowe-Carraco, 93; *Williamson Daily News*, September 12, 1919.

37   Chapman, 235, 228, 229. Source for description of other investors: Chapman, 229; Source for the falling away of other coal men: materials in Box 6 of the Stokes Papers, WVRHC.

38   *Williamson Daily News*, November 19, 1920.

39   *Logan Banner*, May 27, 1921. The author has been unable to locate a Mingo County newspaper of that date.

40   The union had briefly achieved the goal of shutting down the mines at the beginning of the strike in July 1920. By importing replacement workers and enticing men back to work with inflated wages, the operators returned the field to almost full production by the date of the attack on the Himler plant; *Williamson Daily News*, November 26, 1920; J. M. Tulley to Governor John J. Cornwell, May 26, 1920, Cornwell Papers, WVRHC; *Williamson Daily News*, December 14, 1920.

41   *1908–1923 Annual Reports of the West Virginia Department of Mines*; Chapman, 235–237; *Logan Banner*, May 27, 1921. The 1921 run of the *Williamson Daily News* was not available.

42   *1914–1917 Annual Reports Department of Mine Reports*.

43   *Mingo Republican*, March 1, 1917. According to the article, two-thirds of the 30 percent increase had occurred in a single year between January 1916 and 1917. The inflation of food prices in Mingo County inspired "old timers" to compare their current situation to that of the 1860s, during the Civil War; Author's extrapolation based on the comparative rise of food prices and wages in Mingo

County. Food price figures taken from the *Mingo Republican*, and wage rates from *Annual Reports of the Department of Mines*.

44   Keith Dix, *What's A Miner to Do: The Mechanization of Coal Mining* (Pittsburgh: University of Pittsburgh Press, 1988), 205; Munn, 251.

45   Munn, 251; "Testimony of W. E. Hutchinson," *West Virginia Coal Fields*, 79.

46   Munn, 251 and Shifflett, 67; Shifflett, 41, 58; *Williamson Daily News*, December 24, 1919.

47   Munn, 245.

48   *Mingo Republican*, August 16, 1917; Paul Salstrom, *Appalachia's Path to Dependency: Rethinking a Region's Economic History 1730–1940* (Lexington: University Press of Kentucky, 1994), 62.

49   J. H. Stewart, Commissioner of the State of West Virginia Department of Agriculture, *Third Biennial Report of the West Virginia Department of Agriculture, 1917–1918* (Charleston, WV: Tribune Printing Company, 1918), 37, 37–38, 39.

50   *Mingo Republican*, September 27, 1917.

51   *Williamson Daily News*, November 19, 1919.

52   Hennen, *Americanization of West Virginia,* 77–79.

53   *Williamson Daily News*, April 4, 1918.

54   *Williamson Daily News*, April 4, 1918.

55   Munn, 247. The same investment consortium that owned and operated Logan's Island Creek Coal company backed the Pond Creek Coal Company which operated across the Tug Fork River from Williamson. The U.S. Steel subsidiary, U.S. Coal & Coke, through local management, controlled the Crystal Block Coal & Coke Company; the N&W operated captive mines at Vulcan and Chattaroy.

56   *Williamson Daily News*, January 4, 1919. Both Solvay and Island Creek entrusted their company magazine projects to veteran newspaperman George C. McIntosh. McIntosh Memoir, WVRHC.

57   For examples of the amenities which the Williamson-Thacker operators claimed to provide for their employees, see: *The United Mine Workers in West Virginia*, 25; and Floyd Bunyon Shelton, "An Investigation of the Social Life of a West Virginia Coal Field," Bachelor's thesis, Emory University, 1920; Contemporary accounts of the 1920–1922 strike, including local newspapers, state police reports to Governor Cornwell, and several subsequent Matewan Oral History Project narratives assert that strike "hotspots" included Sprigg, Chattaroy,

Pond Creek, and New Howard, all of which were communities dominated by highly capitalized mines.

58   Most of West Virginia's initial wartime legislation was passed during the second extraordinary session of 1917, from May 14–26. *Acts of the West Virginia Legislature* (Charleston, WV: Tribune Printing Company, 1917); Hennen, 37.

59   Chapter 9: "Special Deputy Bill," House Bill #34, *Acts of the West Virginia Legislature for the year 1917* (Charleston, WV: Tribune Printing Company, 1917), 46–47; also reported in *Mingo Republican*, June 7, 1917. The special deputies law was passed over the opposition of West Virginia's Federation of Labor. Frank W. Snyder, the editor of the *West Virginia Federationist* denounced both the special deputies' law and the vagrancy (Slacker) law as weapons which could be used to suppress labor organizations in the state. Snyder cited as evidence the alleged strikebreaking activities of special deputies in Boone, Braxton, and Logan counties. *West Virginia Federationist*, August 4, 1917, quoted in Merle T. Cole, "The Department of Special Deputy Police, 1917–1919," *West Virginia History* 44 (Summer 1983): 321–333; Until the 1910 U.S. Census, Williamson had been the center of Lee District; the city's rise in population validated Williamson becoming its own district. *Abstract of the Thirteenth Census, with Supplement for West Virginia*, 577; *Mingo Republican*, July 19, 1917.

60   Four newspaper articles from the *Williamson Daily News* tell the story of this incident: *Williamson Daily News*, August 22, 1918, August 24, 1918, March 5, 1919, and April 17, 1920; The "Home Defense Guards" were a unit from Huntington that was one of four units started in the state; the other three were in Point Pleasant, Morgantown, and White Sulphur Springs: "West Virginia in the War," 786; Breeden is located in the same district as Dingess. Despite the lawless reputation of this section, very few special deputies were posted there, instead most served in Mingo's heavily developed coal districts; Source for number of draft dodgers from Harvey district: *Mingo Republican*, August 23, 1917.

61   *Mingo Republican*, August 23, 1917.

62   *Williamson Daily News*, April 17, 1920; Perry, 49.

63   Chapter 12, "The Idleness and Vagrancy Act," Senate Bill #7, in *Acts of the West Virginia Legislature, Second Extraordinary Session*, May 14–26 (Charleston, WV: Tribune Printing Company, 1917), 51–53; 1917 *Legislative Handbook and Manual of the State of West Virginia*, 463; Hennen, 47; West Virginia State Council of Defense, *Report of the Secretary on the Operation of the Compulsory Work Law, for*

the Year Ending June 19, 1918. (Charleston, WV: Tribune Printing Company, 1918), 7, Pamphlet 2902, Pamphlet Collection, WVRHC.

64   "West Virginia in the War," 791–792.

65   Diner, 236; Corbin, *Life, Work, and Rebellion,* 177, 180, 184, 198; Johnson, 82. The New River (smokeless) field was the only southern West Virginia anti-union field organized during the war.

66   Johnson, 97; Corbin, *Life, Work, and Rebellion,* 181.

67   Dix, *Work Relations,* 205; Corbin, *Life, Work, and Rebellion,* 197; Johnson, 97.

68   Mooney, 59–60; Barkey, 188; Corbin, *Life, Work, and Rebellion,* 184–188; See author's dissertation chapter six for the contributions of West Virginia's anti-union coal operators.

69   Johnson, 211; The old leaders, who had been driven from power following the Paint and Cabin Creeks strikes were gearing up to challenge the insurgents in District 17's pending election. John J. Cornwell to N. A. McKenzie, September 24, 1920, and N. A. McKenzie to John J. Cornwell, December 3, 1920, Cornwell Papers, WVRHC.

70   *Williamson Daily News,* June 29, 1919.

71   *Williamson Daily News,* June 29, 1919; Johnson, 95–96.

72   *Williamson Daily News,* June 29, 1919. See Hennen, *Americanization* for more on the sinister goals behind Cornwell's platitudes.

73   *Williamson Daily News,* June 29, 1919; *Williamson Daily News,* August 18, 1919.

74   Mooney, 59–71; Source for Chafin: Lee, 92–93; Mooney, *Struggle in the Coal Fields,* 63–64; John L. Spivak, *A Man in His Time* (New York: Horizon Press, 1967), 58.

75   Fisher, 377; Corbin, *Life, Work and Rebellion* and Fagge *Power, Culture and Conflict in the Coalfields,* offer more extensive analyses of the "First Armed March on Logan."

76   Spivak, 69–70; Corbin, *Life, Work, and Rebellion,* 100; Fisher, 377.

77   Fisher, 378–379; *Williamson Daily News,* September 8, 1919.

78   John L. Lewis to Governor John J. Cornwell, September 11, 1919, quoted in Fisher, 380.

79   *Williamson Daily News,* September 20, 1919; Johnson,101; *Williamson Daily News,* September 13, 1919.

80   Johnson, 102–103.

81   Johnson, 103; Fisher, 379; Hennen, *Americanization*, 93.

82   Johnson, 98–102.

83   Corbin, *Life, Work, and Rebellion*, 197; "The West Virginia Miners: Extension of Remarks of Honorable Wells Goodykoontz, of West Virginia in the House of Representatives," Saturday, November 1, 1919: *Congressional Record*: 8349–8350, 8350. Someone, probably Goodykoontz, sent a copy of the speech, which had been printed in the *Congressional Record*, to S. D. Stokes, Stokes Papers, WVRHC.

84   Johnson, 98–104.

85   "Testimony of Fred Mooney," *West Virginia Coal Fields*, 15; *Williamson Daily News*, November 19, 1919; According to Cornwell biographer Lucy Lee Fisher, during the general strike, non-union miners produced 75 percent of the state's normal output. Fisher, 380; G. T. Blankenship to John J. Cornwell, telegram, November 11, 1919, Cornwell Papers, WVRHC.

86   C. M. Gates to John J. Cornwell, November 19, 1919, Cornwell Papers, WVRHC; The cases referred to are the Vulcan Collieries night school program and the founding of the Pond Creek Improvement Club, a voluntary association of operators, miners, and private citizens. *Williamson Daily News*, November 19, and November 14, 1919; Gates to Cornwell, Cornwell Papers, WVRHC.

87   "Goodykoontz House of Representatives speech," Stokes Papers, WVRHC.

88   *United Mine Workers' Journal* 30 (July 30, 1920): 8.

89   *United Mine Workers Journal* 32 (November 15, 1921):10; Shifflett, 41.

90   Figures extrapolated from the 1917, 1918, and 1919 *Annual Reports of the West Virginia Department of Mines.*

91   Figures extrapolated from the 1917, 1918, and 1919 *Annual Reports of the West Virginia Department of Mines; Dix, What's a Miner to Do*, 141.

92   Until 1912, Mingo ranked third among West Virginia's counties that had sent inmates to the West Virginia state penitentiary. By 1919, Mingo sent fewer than any other coal-producing county, but at the war's height in 1918, there were 11 murder indictments in the county. "Judge James Damron" entry in George W. Atkinson, *Bench and Bar of West Virginia* (Charleston, WV: Virginia Law Book Company, 1919), 354. Source for number of murders: Lane, 100; Matewan and Thacker, in particular, were combination coal and independent towns. In both, the saloons operated by elites like M. Z. White and R. W. Buskirk catered to the min-

ers who were an important "constituent" base for White and Buskirk. Buskirk's departure from Matewan had been heralded as an opportunity for the "good citizens" of the town to build a new reputation for the community: *Mingo Republican,* April 29, 1915; Elihu Boggs, although a miner by profession, served as clerk of the county court from 1914–1920. Similarly, at least one Matewan Massacre defendant had served as a deputy sheriff. See author's dissertation chapters six and ten for details on these examples.

93   Readers are reminded that several leaders of the "war effort" in Mingo, including Randolph Bias and Hiram S. White, had, throughout the period before the war, fought repeatedly against the Hatfield Machine. See author's dissertation chapters three and six; *Williamson Daily News,* March 18, 1914; *Williamson Daily News,* June 21, 1919; "Testimony of C.E. Lively," February 25–26, 1921, unknown newspaper Matewan Omnibus Collection, ERCA; *Williamson Daily News,* May 15, 1920. See author's dissertation chapter 10 for more on this issue.

94   *Williamson Daily News,* October 13, 1919; "Olmsted Exhibit No.3," in "Testimony of Harry Olmsted," *West Virginia Coal Fields,* 223–271: 230–236 and *Williamson Daily News,* October 13, 1919. "Olmsted Exhibit No.3" contains a list of all of the registered members of the Mingo Militia and which community they resided in. For the members from Williamson, individual occupations were also listed.

5

THE MASSACRE: BEFORE & AFTER

"The Worst Has Come"[1]
– *Williamson Daily News*

DURING THE EARLY MONTHS of 1920, a series of local, state, and national events began a slow implosion that resulted in tragedy on the streets of Matewan on May 19. The external causes of the terrible tragedy included the uncertain economic state of the coal and rail industries as federal wartime regulation finally ended; the escalation of tension between the UMWA and southern West Virginia's coal operators; and the political contests that played to and focused the discontent of various voting groups. However, the Matewan Massacre also simultaneously vented long-brewing local political, economic, and social resentments. Fully understanding the Matewan Massacre demands a re-examination of its causes, its outcome, and its impact.

Although 1920 later proved to be another banner year for national coal production, in the early months of the year, sluggish economic conditions complicated by inadequate railcar supplies fostered tension in Mingo County. Political tensions had also begun, with early preparations for the general election in November. The Republicans hoped to recapture control of the county government, while Democrats aspired to retain their primacy. In scrambling for ways to improve their respective positions, the local coal and political leaders of Mingo made decisions that directly influenced their subsequent response to the violence in Matewan on May 19, 1920. Other tragedies preceded the Massacre, making already edgy nerves even easier to trigger. At seven o'clock Monday morning, January 12, 1920, an explosion rocked the city of Williamson. The blast shook every building within a mile radius of the city and shattered every window from the plate glass of the storefronts to those of the private residences. As the frightened and confused citizenry wandered the streets, "fragments" of the body of James Childers, the foreman of the Superior-Thacker Coal Company, were found throughout the city. It was soon learned that the company's magazine building, which had contained ten cases of composition mining explosives, had exploded shortly before the company's miners had lined up to receive their daily allotment of the material. Although relieved that the blast had claimed just one casualty, the citizens of Williamson were outraged that such a danger had lain "almost in the heart of the town." How and why the explosion occurred was never clearly established. In the months that followed, Superior-Thacker refused to accept responsibility for the incident. Citing economic hardship, the company offered the citizens of the city a settlement which allowed Superior-Thacker to repay only 25 percent of the cost of the property damage. Most of the citizens who had actually pursued claims against the company accepted the terms.[2]

The Superior-Thacker explosion symbolizes the status of the coal industry and labor relations in Mingo County in the months before the Matewan Massacre. Generally poor employment, production, and transportation conditions, combined with the escalating conflict between the union and the operators, did not present any new danger that would have alerted observers to the potential for the raw violence of the Massacre. Familiar with the vicissitudes of the economic situation and labor relations rhetoric, townspeople did not foresee the danger until the fateful blast. However, an examination of the status of the coal industry nationally and in southern West Virginia in early 1920 reveals how it provided the backdrop to the events that occurred on the streets of Matewan on May 19.

In the late winter and early spring of 1920, resolving the leftover tensions from the national coal strike of 1919 preoccupied the federal government. Despite the general effectiveness of wartime federal regulation, the fitful and uneven dismantling of government controls created a climate of resentment and distrust. While the government had released the coal industry from price controls and distribution directives, it denied the UMWA request for a wage increase with a specious semantic argument over whether the war had actually ended. On January 22, 1920, when the chairman of the Senate Committee on Education and Labor introduced a bill to create a National Labor Board that would have extended the federal role in labor relations, the measure received little support or attention. Both the UMWA and the leaders of the coal industry were too busy shoring up their own positions and presenting their case to the Bituminous Coal Commission, the federal investigative body created as part of the settlement of the 1919 strike.[3]

John L. Lewis. *In January 1920, Lewis "declared war" in the effort to unionize southern West Virginia. [WVRHC]*

While UMWA representative Van Bittner testified before the commission that miners' earnings had lagged far behind both the inflation of the cost of living and the wages paid to other industrial workers during the war, acting union president John L. Lewis traveled to West Virginia to announce that the UMWA was launching an organization drive that would, once and for all, unionize southern West Virginia. Lewis had foreseen the potential of a unique situation. First, it was highly probable that the Coal Commission would find that union miners deserved a wage increase. Second, Lewis would also have seen that given the prevailing economic conditions in southern West Virginia, the miners employed in the non-union coalfields were likely to clamor for the same concession. By renewing the union's pledge to organize southern West Virginia before the Bituminous Coal Commission issued its report, John L. Lewis positioned the union to materially benefit from both the federally sanctioned wage increase and the discontent rampant in southern West Virginia at the time.[4]

During the late winter of 1919–1920, national coal production had been curtailed at least 25 percent. In the Williamson-Thacker coalfield, production fluctuated between 30 and 60 percent, well below the national average. As discussed earlier, a major contributing factor to the dismal coal situation was a transportation crisis. The national rail systems were still under federal regulation, and there were not enough cars to move southern West Virginia's coal. Those hardest hit by the prevailing conditions were the miners. Working fewer and shorter days, the miners faced increasing tangential pressures that heightened resentment against their employers. These tensions worsened when shortly after a neighboring coalfield reported an outbreak of influenza, some companies raised the miners' compulsory payment for medical services.[5]

While struggling to survive the closing days of winter with less income and high living expenses, miners in the Williamson-Thacker coalfield watched helplessly as their employers belligerently defied local governments. The Hunt-Forbes Coal Company refused to compensate the city of Williamson for the damage its coal hauling had inflicted on the city streets. In Kermit, the Gray Eagle Coal Company concluded a three-year fight with the town council and the county court over a public road by hiring guards to protect company property on the contested land. As the actions of these companies illustrated, the coal companies of the Williamson-Thacker coalfield either blindly or arrogantly disregarded the potential impact of their actions.[6]

In addition to imposing higher prices for company services and yet sidestepping their own local financial responsibilities, the coal companies engaged in speculative ventures that further destabilized employee relations. In January, M. A. Hanna & Company purchased one of the largest companies in the coalfield, the Red Jacket Consolidated Coal & Coke Company. Less than two months later, Red Jacket reportedly went back on the block. Although accounts of the pending sale proved false, the company's approximate sale value of two million dollars was revealed. Moreover, another company, the Buffalo-Thacker Coal Company purchased a mining operation in another coalfield on the Little Coal River. These and other coal business deals convinced the miners of the Williamson-Thacker coalfield that not everyone was suffering from the hard times.[7]

Between February 6 and April 16, 1920, a series of seemingly disconnected events underscored why the miners of Mingo County responded, after nearly twenty years' indifference, to the union's call. Taken together, the events illuminate why the miners came to believe that their best hope for economic security and political freedom lay in joining the UMWA.

First, just one week after John L. Lewis "invaded" Bluefield, Governor Cornwell addressed the West Virginia Lumber and Builders' Supply Association in the same city. While claiming to be a trade unionist, the governor praised the role of southern West Virginia's non-union operators in breaking the 1919 strike. Cornwell went on to explain that if the union succeeded in organizing these same non-union coalfields, nothing would stop the nationalization of the coal industry. When Cornwell pledged to resign if he failed to keep the peace during the pending effort, his audience leaped to its feet in a standing ovation and shouted its support. Deliberately or not, the governor had exposed whose ally he would be.[8]

Any lingering doubts the miners might have had about Cornwell's position were dispelled by how the hearings of the governor's commission investigating the 1919 Armed March on Logan ended just one month later, on March 13. Headed by Governor Cornwell's hand-picked proponents of law and order, the commission had lost the cooperation of District 17 officials and concluded not with substantive recommendations for addressing the endemic problems of labor relations in southern West Virginia, but with a blanket condemnation of the miners' actions and those of District 17's leadership in particular. The miners of southern West Virginia were forced to acknowledge that their governor had not only allied himself with their employers, but had no interest in giving their grievances an impartial public hearing.[9]

The public announcement on March 29, 1920, that the Bituminous Coal Commission had recommended that coal miners receive a 27-percent wage increase only underscored the benefits of union membership. To add insult to injury, one of Mingo County's two newspapers, the *Williamson Daily News*, printed an editorial proclaiming coal a mismanaged industry, that a wage increase for miners was "only right," and that the miners should not have to suffer because of "the mismanagement of others." Less than a week after the *Daily News*'s apparent statement of empathy for the miners' cause, strikes broke out among the railroad workers of the Norfolk & Western.[10]

Adding to the building pressure these incidents created, there was one more culminating event that drove the miners of Mingo to action. In late March or early April 1920, the miners at the Howard Colliery at Chattaroy asked for a ten-cents-per-car raise to offset an increase in the cost of living. After calling in his superintendent from Bluefield, the Howard manager asked the men to give them a week to consider the request. He also asked the men to continue working, which they did. Three days later the miners of the Howard Colliery were offered a nine-cent raise, to which they agreed and were "well satisfied." At the end of their shift that same day, however, the men exited the mines only to find a notice that the prices of all their necessary work articles had been raised five to twenty-five cents; more-over, it would now be compulsory that the miners buy exclusively from the Howard Colliery company store. The miners of the Howard Colliery had negotiated in good faith and felt betrayed.[11]

On the morning of April 16, the miners of Burnwell Coal & Coke arrived at the mines and found the following sign posted on the drift mouth:

> To the miners of Burnwell Coal Company: We shall have this 27 per cent raise; we want this 27 per cent raise which the Government has granted us.

The Burnwell miners refused to enter the mines. Eighty of the ninety-two Burnwell employees signed a letter and sent two of their own to Charleston to request a charter from District 17. When asked why they undertook this action, one of the men replied that it was because the miners wanted "to belong to an organization of [their] own craft." District 17 Secretary-Treasurer Fred Mooney sent the Burnwell miners back to Mingo with a promise to send organizational assistance if the men returned to work. The Mingo organization drive had begun.[12]

Within days, large gatherings of miners were orchestrated. Miners flocked by the hundreds to take the union obligation. On April 22, approximately 500 men attended a meeting at the Baptist Church in Matewan, and between 275 and 300 joined the union. The next day, also in Matewan, between 700 and 800 miners congregated. The response seemed to portend that the miners of Burnwell would get their wish. For four heady days, W. E. Hutchinson and the other employees of the Burnwell Coal & Coke Company went into the mines as union men. However, on April 27, dismissal notices were delivered, giving the union miners and their families just three days to vacate company housing. Their employer, Mr. Pritchard, declared that "he would let his mine go until moss grows over it, until it falls into the huckleberry ridge, before he would ever work a union man." The battle line had been drawn.[13]

Shortly after the organizational flurry began, the coal companies of the Williamson-Thacker coalfield launched their effort to stem the tide. The day the Burnwell miners received their eviction notices, the *Williamson Daily News* reported that it had "talked to both sides, but got no hope as to a reasonable settlement." The *News* pleaded with both sides to "remember that people will suffer" and to "keep within the bounds of the law." Just two days later, on April 29, reports of operator activity revealed that the *News*'s appeals had fallen on deaf ears. The companies began evicting the men who were assisting in the organization effort. The operators also imported detectives, machine guns, and high-powered rifles. Although the operators would later deny any culpability in the escalation to violence, a single item from the same newspaper article illuminated the first clue to the immediate cause of the Matewan Massacre, now less than three weeks away.[14]

On April 27, 1920, Mingo County Sheriff G. T. Blankenship arrested Albert C. Felts, brother of the head of the Baldwin-Felts Detective Agency, for illegally processing evictions. Felts posted a $2,000 bond and appeared in magistrates' court two days later, where he and 27 other agents were placed under a peace bond. Hoping to de-escalate the situation, Sheriff Blankenship called a meeting with miners before the courthouse in Williamson. Blankenship asked the miners if they would peacefully comply with their evictions if he and his deputies processed the writs. According to Blankenship, the miners agreed. Although Blankenship later claimed to have processed hundreds of evictions for the coal companies, less than two dozen were undertaken between mid-April and May 19.[15]

There were at least two reasons for the low number of evictions. First, despite a promise to process any evictions lawfully ordered by the circuit court, Blankenship in fact engaged in stalling tactics, asking companies to obtain both ten-day notices from the court and then three-day notices from justices of the peace. Second, in early May, after holding a strategy session with Baldwin-Felts Detective Agency Chief Thomas L. Felts, the operators began compelling their employees to sign "yellow dog" contracts. The contracts not only precluded union membership but also contained a clause that must have been particularly galling to many miners. In addition to agreeing not to join the union, the men were also required to endorse a denunciation of the union. At least two of the companies within the vicinity of Matewan, the Stone Mountain Coal Corporation and the Red Jacket Consolidated Coal & Coke Company, instituted this policy in early May.[16]

Despite the evictions and the imposition of the "yellow dog" contracts, the organization of Mingo's miners continued unabated during the first two weeks of May. Emboldened by Sheriff Blankenship's vow to protect them from unlawful acts by the companies, the miners held mass public meetings. Three thousand attended a union meeting in Matewan on May 6, 1920. Eight days later, 200 miners took the union pledge at Williamson. On that same day, May 14, another meeting was held in Matewan, although no organizers or representatives of the UMWA were present. The speakers included Hugh Combs, a miner and Southern Methodist "exhorter"; an African-American minister identified only as "Johnson"; and George Allen, a merchant from Thacker, all of whom endorsed the union as the miners' advocate. Combs informed the miners that joining the union was a just act of self-protection against the operators who were themselves organized. The African-American minister informed his fellow black miners that Abraham Lincoln had given them their freedom, but now the union would "give them their liberty." Allen told those assembled that he had once been a miner and, because he was in sympathy with their cause, pledged the use of his store if the miners needed a place to meet. That the miners spoke so freely because of Blankenship's protection and had an open forum in his hometown of Matewan soon also received public acknowledgment. During the annual meeting of the West Virginia State Federation of Labor, District 17 Secretary Fred Mooney informed his fellow conventioneers that "the UMW had the support and backing of the Mingo County officials, a condition which had never existed before in the history of the organization." Reported in the Bluefield Daily Telegraph, Mooney's comments further galvanized the anti-union sentiments of Mingo's coal operators.[17]

While Fred Mooney celebrated the establishment of the union beachhead in Mingo, the *Williamson Daily News* made increasingly ominous observations about conditions in Mingo County. The May 8 issue noted, "There is a general restlessness among our laboring people . . . . What will be the outcome of this discontent no one can predict." Just six days later the *Daily News*, after reporting on yet another miners' meeting, noted that "the operators have made nothing public . . . as to what position they will take . . . . Secrecy seems to prevail on both sides."[18]

Just what the *Daily News* expected the two opponents to reveal is not clear. The operators' intractability concerning the UMWA had been openly declared. In addition to Pritchard of Burnwell, other companies publicly advertised their position. On May 5, Borderland posted a notice that paralleled Pritchard's comment in clarity. The notice read, "This is a free country . . . but . . . no union men shall be employed by this company." Stone Mountain posted a similar notice on the front window of its company store in Matewan.[19]

For its part, District 17 and the national leadership of the UMWA also moved forward with the formalities of its organization effort. After the state Federation of Labor convention ended, the first international organizers arrived in Mingo. By Monday, May 17, 1920, the union had initiated efforts to set up a tent colony near Matewan for the evicted union miners. District 17 official C. H. Workman arrived that day with tents and instructions to lease all land available for a tent colony. Just two days later, on Wednesday, May 19, Albert and Lee Felts, leading a contingent of Baldwin-Felts agents, stepped off the train at 11:47 a.m. Their assignment—to ensure the eviction of several families from Stone Mountain company housing in the town of Matewan. Less than six hours later, the Felts brothers, five of their men, Matewan's mayor, and two bystanders would lie dead or dying on the streets of the town. However large a role it may have played in this tragedy, economic oppression is not sufficient unto itself to spark an encounter like that which occurred in Matewan on May 19. Other studies of violence in Appalachia corroborate this.[20]

As 1919 drew to a close and the political season of 1920 opened at the national, state, and local levels, Republicans dominated the stage. After orchestrating the defeat of President Wilson's internationalist agenda (personified by the League of Nations), and driving him to utter physical collapse, the nation's leading Republicans turned on each other. At issue was the direction an internally focused nation would follow. The leading liberal contender for the presidential nomination was Hiram Johnson, a Progressive, railroad trust-busting senator from California whose chief allies were Senators William E. Borah and William S. Kenyon of Iowa. Although Kenyon, leader of the United States Senate's "agricultural bloc," was sympathetic to the complaints of organized labor, he had openly denounced civil disorder as a means of addressing these issues. The party's conservative wing backed United States Army General Leonard Wood, a law-and-order advocate. In part because both Johnson and Wood claimed to be the political heir of Theodore Roosevelt, the 1920 Republican national convention deadlocked, resulting in the nomination of Warren G. Harding. Despite their loss, the election-year politics of Wood, Johnson, and Kenyon still affected the labor-dominated politics of West Virginia. Wood, who drew substantial support from southern West Virginia's Republican coal elite, concluded his primary stumping in Williamson, Mingo County, on May 22, 1920, just three days after the Massacre. Both

SAMUEL B. MONTGOMERY

Samuel B. Montgomery. *Pro-labor contender for the 1920 Republican gubernatorial nomination. ["Progressive West Virginian," WVRHC]*

Johnson and Kenyon later served on the Senate's committee assigned to investigate conditions in the West Virginia coalfields in 1921.[21]

Apparently united in their determination to regain the governorship after Cornwell's anomalous tenure, West Virginia's Republican Party suffered from a schism that mirrored the national conflict. Samuel B. Montgomery, a former leader of the West Virginia State Federation of Labor, UMWA attorney, and commissioner of labor under Governor Hatfield, had the support of the liberal wing of the party and the laboring classes of the state. Judge Ephraim F. Morgan

Ephraim F. Morgan. *Northern West Virginia lawyer and business candidate for the 1920 Republican gubernatorial nomination. [WVRHC]*

of north-central West Virginia had the support of the conservative wing of the party, except for the ultraconservative coal elite of southern West Virginia, who supported Colonel Paul Grosscup, an oil and gas executive from Charleston. The three-way contest among Montgomery, Morgan, and Grosscup revealed that the factional legacy of Henry D. Hatfield's gubernatorial tenure not only had persisted, but had expanded. More important, two of the counties soon to be swept up in the crisis between the UMWA and West Virginia's non-union operators, Mingo and McDowell, were the focal point of this state-level political controversy.[22]

The intrafactional fight did not play out in smoke-filled backrooms, but on the front pages of southern West Virginia's newspapers. On January 18, 1920, the *Bluefield Daily Telegraph* revealed the first indication that disharmony ruled in the southern coal counties. According to the *Telegraph*, Mingo Republican party chairman, M. Z. White, issued an endorsement letter for Morgan, while his counterpart in McDowell, Colonel Edward O'Toole, had done the same for Colonel Grosscup.[23]

In the weeks before the primary election, White utilized traditional ma-

FRED PAUL GROSSCUP.

Paul Grosscup. *Candidate of the ultra-conservative wing of the Republican party for the 1920 Republican gubernatorial nomination. ["Progressive West Virginian," WVRHC]*

chine tactics to ensure Mingo's support for Morgan. The March 9 issue of the *Williamson Daily News* asserted that White convened the county's nominating convention by reading a list of candidates he hoped would be considered. When a smattering of objections was raised, White "angrily" adjourned the meeting without considering alternative candidates. The *Daily News* concluded its report by noting that what remained in question was whether "the better citizens in the Republican party . . . resented their continued sale and delivery" enough to act.[24]

Led by Edward O'Toole and T. E. Houston, two of the most powerful coal executives in the smokeless coalfields, McDowell's political factions turned to the West Virginia Supreme Court of Appeals. Less than three weeks before the primary, the Court issued a mandamus ruling that ordered both the unseating of McGinnis Hatfield (a Grosscup partisan) as county chairman and the selection of the Morgan list of election officers. While these actions appeared to indicate politics as usual for southern West Virginia, the fight over the Republican gubernatorial bid released two new forces that profoundly influenced events in southern West Virginia for the next two years.[25]

For the first time, West Virginia's laboring classes, especially its coal miners, had a legitimate advocate-candidate, and their votes could swing the gubernatorial election. The regular party machinery's infighting over Morgan and Grosscup meant that the deciding bloc of votes belonged to Montgomery. The anticipated loyalty of the "organized faithful" inspired what the *Williamson Daily News* referred to as an "entente cordiale" between Grosscup and Montgomery. But in southern West Virginia, the single largest groups of potential Montgomery supporters, the miners, were *not* organized.[26]

It was at this point that the simmering political and labor unrest in southern West Virginia coalesced in Mingo County. According to S. D. Stokes, a Williamson Democrat, it was the three-way fight between Morgan, Montgomery, and Grosscup that actually spawned District 17's effort to unionize southern West Virginia, starting with Mingo. Stokes wrote to a friend that Mingo's Republican chairman, M. Z. White, with the collusion of Greenway Hatfield, had sold the county's primary returns to Morgan supporter T. E. Houston for fifty thousand dollars. However, because White could not keep quiet and publicly proclaimed Mingo for Morgan, Grosscup's campaign manager visited Montgomery's campaign manager in Charleston and reminded him of two important things: first, that Montgomery could carry southern West Virginia if the miners were free to vote their consciences; and second, that the way to "liberate" these voters was to organize them. Thus, Stokes claimed, "In about ten days, the coal operators in this field observed unrest among their laboring men and woke up to the fact that the ordinary labor agitator was . . . abroad in the land." What prevents dismissal of Stokes' version of events as partisan gossip is a review of the timing of White's public actions and the launch of District 17's organization drive. White's letter of support for Morgan had appeared in the January 18 issue of the *Bluefield Daily Telegraph.* Thirteen days later, John L. Lewis announced that the UMWA would initiate a unionization offensive in southern West Virginia.[27]

Mingo County Republican Party Chairman and State Senator, M. Z. White. *White and his wife were pulled to safety during the Massacre by African-American businessman John Brown. [WVRCH]*

The turmoil in Republican ranks in McDowell and Mingo counties extended into their most loyal constituency—the African-American community. On the same day that the *Bluefield Daily Telegraph* revealed the schism between the leaders of the party, it also noted that a "New Emancipation Movement" had been started by the "Colored Republican Laboring Men's Organization of McDowell County." The avowed intent of the organization was not hostile to the operators or any other business interests but had as its only goal "to make an effort to end negro political slavery." By May, the movement had spread to Mingo County. At a meeting held at the county courthouse in Williamson on May 14, the African-American voters of Mingo County were exhorted to vote independently because the "Republican bosses hang on their necks at election time" but promptly forget them after. As the *Williamson Daily News* observed, the speakers and their audience also spoke enthusiastically in favor of the organization efforts of the UMWA.[28]

Besieged by the union organizers, who also stumped for Montgomery, and threatened with the defection of the black vote, Mingo's Republican machine intensified its efforts to regain control over the pending primary election. County chairman White sued the Democrat-controlled county court because it required the list of registered voters in Mingo to exclude voters whose registrations were disputed by precinct officers. The West Virginia Supreme Court of Appeals granted White a *writ of mandamus* on May 18, 1920, one day before the Matewan Massacre.[29]

Although the report of the supreme court of appeals does not reveal that the disputed registrations included any voters in Matewan's precinct, there is evidence that indicates how the town came to be the focal point for the political and labor agitation in Mingo. First, in the spring of 1920, the county's two most powerful officials, Sheriff Blankenship and the president of the county court, E. B. Chambers (Blankenship's brother-in-law), were from Matewan. Second, the town's mayor and Chambers' ally, C. C. Testerman, was targeted for criminal prosecution shortly after winning a second term. The statement Testerman issued after being found not guilty of prohibition violations reveals not only the town's political tensions, but also his determination to oppose the forces arrayed against him. Testerman asserted:

> I am mayor of Matewan, elected by the people for a second term, which speaks for itself. They can say what they please, law when

The Matewan Massacre defendants, ca. 1921. *Members of the Chambers family included: Reece, Ed, and Halley Chambers, and Chambers in-law Clare Overstreet. [WVRHC]*

they please, fight when they please, but I am going to run the town according to the laws of this state and the best of my ability. Instead of our town having the name it has possessed these many years, you are going to hear of Matewan representing the highest order of good government. Our elections are held quietly now. We have no drunks, and our people at last can go to church unmolested and worship as they please. I intend to do what is right the best I know how and the threats of evil doers won't alter me in my course one inch.

Under Testerman's tenure, during the early spring of 1920, Matewan became a free assemblage haven for the organization of Mingo's miners. Testerman's allies, Blankenship and the county court, also provided protection for the union cause. Half of the county's miners lived in the Magnolia District, of which Matewan was the center. To effectively disenfranchise these voters, the protective power of the Matewan Democrats had to be neutralized. The Matewan Massacre occurred six days before the primary election was held.[30]

When the primary was held, the newspapers reported on the unnatural state of quietness and the low voter turnout. District 17 officers fumed to no avail about the number of ballots floating in the river. Although he lost the state by more than 160,000 votes, pro-labor candidate Montgomery lost Mingo County to Morgan by 15 votes in the Republican gubernatorial primary race. When the Matewan Massacre trial ended in acquittal in March 1921, four of the remaining 16 defendants were members of the Chambers family of Matewan, a fact that had led at least one Hatfield family descendant to refer to the Massacre as "the Chambers' War."[31]

On May 19, Matewan's Chief of Police Sid Hatfield and a group of deputized citizens and miners engaged in a gun battle with representatives of the Baldwin-Felts Detective Agency. Ten men, including Mayor Testerman died as a result. Within weeks, the Operators' Association of the Williamson-Thacker coalfield initiated a lockout in an attempt to defeat the rampant success of the UMWA's organization drive. In retaliation, the union issued a strike call, which it rescinded only after 28 months of bloody conflict and violent oppression, including the retaliatory assassination of Hatfield and the Battle of Blair Mountain, the largest armed civilian insurrection since the United States' Civil War. History has recorded the impact of this second "mine war" on the future of the United Mine Workers of America in West Virginia. What has been overlooked is the local effect of the Massacre and subsequent events, as well as how the story of those events was transformed to serve the agendas of people and interests far from Mingo County.

In the space of a few minutes, the event that became known as the Matewan Massacre transformed the social, economic, and political relations of Mingo County. Prior to May 19, 1920, class or occupation had not been the determining factor in an individual's access to political power or social status. However, how the local citizens viewed the Massacre reconfigured the course of their public associations. To the striking miners and their allies, the Massacre was a glorious instance of retributive justice—Sid Hatfield and his compatriots had given the Baldwin-Felts agents what they had coming. For other members of the community, the incident did not reflect the "identity" of Matewan, and as such was a source of outrage and shame. As one contemporary reporter observed, "The difficulty has now become a bigger proposition than any ordinary strike . . . . It is the ranging of a community into opposing factions."[32]

The charitable organizations, whose local and state chapters were dominated by pro-industry elites, refused to offer the striking miners and their families any relief. Both the Red Cross and the local chapter of the YMCA refused humanitarian aid to the inhabitants of Mingo's tent colonies. The Red Cross denied assistance because "no Act of God" was responsible for the miners' plight, while the YMCA denounced the colonies as centers of immorality. After another "battle" occurred in Mingo shortly before the first anniversary of the Massacre, the local chapter president of the YMCA helped organize a vigilante organization that helped restore "law and order" to Mingo County.[33]

Social affiliations directly influenced and were in turn affected by the Massacre and strike. The Massacre and strike also divided people of faith in Mingo County. Before May 19, 1920, religious affiliation had not automatically characterized an individual's class status. For example, area natives, including miners' families and operators together had founded and attended the Matewan Methodist Church. However, when Church of Christ minister Mose Alley "went bond" for Massacre defendant Reece Chambers, at least one member of his congregation denounced him. Mrs. Mary Duty, on whose porch Agent J. W. Ferguson had been killed, succinctly condemned Alley by stating, "If Preacher Alley will swallow and shield this I am through with him."[34]

The social-religious cleavage over the Massacre rapidly took on class overtones. After the Massacre, one woman remembered, "The churches gave no haven to the poor miners." This comment reflects the social differentiation based on religious affiliation that had been brought into sharp focus by the labor conflict in Mingo. "Mainstream" Protestants, whose churches were led by educated clergy, remained aloof from the situation, with the prominent exception of Williamson's Presbyterian minister, J. W. Carpenter, who condemned the strike from his pulpit and also helped organize the Mingo Militia. By contrast, the miners drew strength and inspiration from their evangelical Christian beliefs. The inhabitants of the striking miners' tent colonies, in addition to singing hymns, used religious imagery to express their commitment to the struggle in letters to the *United Mine Workers Journal*. The centrality of the miners' faith was underscored by the attention religious affiliation received during the Matewan Massacre trial in 1921. When Hugh Combs, the man entrusted by Mayor Testerman to select men to protect Matewan on May 19, testified during the Massacre trial, he was forced to explain his vocation as a Methodist exhorter because prosecuting and defense lawyers argued about whether he was a "Holy Roller." In the wake of the 1920–1922 strike in Mingo, membership in mainstream Protestant churches or Pentecostal and Holiness churches became an automatic indicator of an individual's place in the local social order.[35]

Because it elicited such a divisive reaction from the inhabitants of Mingo County, the Matewan Massacre also transformed the atmosphere of everyday public intercourse. For the first year of the conflict, Sheriff Blankenship's and Sid Hatfield's support for the strike made Matewan the miners' town. Non-cooperative merchants were boycotted. Without fear of official retribution, many miners harassed and intimidated those who failed to support them. Two weeks after murdering an abusive former deputy sheriff and railroad guard, miners erected a mock effigy of his gravesite on a sandbar in the river at Matewan. The slain man's widow also claimed that one of the men acquitted of the murder confronted her on the street, raising her veil and laughing in her face. After the murder of Ance Hatfield, the manager of the Urias Hotel and star witness against Sid Hatfield and the other Massacre defendants, an "exodus" from Matewan began that lasted for more than a year. Those who remained in the town either supported the miners for their own political or material reasons or sought to remain neutral. The merchants who had aided the miners during the strike were rewarded with the miners' patronage for as long as they remained in business. Although some businessmen, such as John Brown, eventually returned to Matewan; others, such as brothers Joseph and Samuel Schaeffer, left forever.[36]

The most profound and lasting impact of the Massacre lay in its effect on local politics. As we have seen, Mingo County had never been governed by a single party funded by absentee industrial interests. In fact, there is also much to suggest that District 17 chose Mingo County to launch the 1920 southern West Virginia organization drive specifically because the union hoped to turn the factional divisiveness of the county's political elite to its own advantage. Moreover, testimony from the Massacre trial suggests that old political animosities might have played a role in the escalation to violence on May 19, 1920. In the Massacre's aftermath, at least initially, the actions of certain Mingo County leaders indicate that the union strategy had succeeded. E. B. Chambers and A. C. Pinson, the leaders of the two Democratic factions in the county, both contributed to the bonds of the Matewan men indicted for the Massacre. Judge Damron, a maverick Republican who won over the miners by overseeing the indictments of local coal officials, guards, and Baldwin-Felts detectives, received the miners' endorsement for his 1920 reelection bid. Even the former chairman of the County Republican organization Greenway Hatfield signed a union contract for his mine.[37]

However, by the early fall of 1920, most of Mingo's leaders had abandoned their pro-union stance. James Damron resigned from the bench and withdrew from the judge's race. By the time the Massacre defendants went to trial in January 1921, Damron had joined the prosecutorial team. A. C. Pinson, after defeating Greenway Hatfield in the 1920 sheriff's race, cooperated wholeheartedly with the efforts of Governor Cornwell and the operators to secure the miners' defeat. One miner later alleged that Pinson had accepted the bribes Blankenship refused, and in return for fifteen thousand dollars, allowed the operators to mount machine guns on the rooftops in Matewan. Even E. B. Chambers evaded public identification with the miners' cause when members of the senate committee traveled to West Virginia on a fact-finding mission in 1921.[38]

After the UMWA abandoned the Williamson-Thacker strike in October 1922, it appeared that the pre-strike political equilibrium of machine rule had been restored in Mingo County. Fellow Williamson City Ring man, Alex Bishop, succeeded Pinson as sheriff in 1924 and was in turn followed by Greenway Hatfield. However, with the long-delayed liberation of the miners by Franklin Roosevelt's New Deal legislation, the pro-union stand taken by G. T. Blankenship and his Chambers' in-laws was finally rewarded. Although Blankenship never reentered Mingo County politics, Thurman "Broggs" Chambers and several other Chambers family members or their allies served terms as Mingo County sheriff. For the next three decades, the politicians who ruled Mingo held power as a result of tactics pioneered by the Chambers' family, or of their own connections to families that had also sided with the miners during the 1920–1922 strike. However, the politicians the miners rewarded with control of Mingo County proved as exploitive as the old Williamson and Hatfield machines. Remarking on the corruption uncovered during a Justice Department investigation of Mingo County politics in the late 1960s, a United States attorney observed, "Freedom has been lost in Mingo County. There is a government of the organization, by the organization, and for the organization." How little times had changed along the banks of the Tug Fork River.[39]

The abandonment of the Williamson-Thacker strike nearly two and a half years after the Matewan Massacre underscores the fatal tragedy of May 19, 1920. Mingo's miners, who had been earning nearly seven dollars for an eight-hour shift in 1920, were, by the mid-1920s laboring for two dollars for a 12-hour day or longer, if they worked at all. In 1924, District 17 was stripped of its autonomy, and union membership among the state's miners fell to a negligible number. Despite John L. Lewis' vigorous and forceful leadership, or maybe because of it, the UMWA nearly collapsed in the decade between the Williamson-Thacker strike and the dawn of the New Deal. By 1925, only six of the mining companies operating in Mingo County in 1910 were still running, and they were all subsidiaries of industry giants such as U.S. Steel. In spite of the success of the anti-union campaign of the early 1920s, the national coal industry, as one historian noted, "went from riches to rags" even before the Great Depression hit in 1929. How the union, the operators, and the government reacted to the Matewan Massacre and Williamson-Thacker strike provides insight into how this state of affairs transpired.[40]

The Matewan Massacre was more than just a deadly encounter between outraged miners and the agents of their oppression. Previous scholars, by "anatomising" the Massacre's historical significance and relegating it to the long struggle for labor rights in West Virginia, have misunderstood not only its impact on life in Mingo County, but also what it illuminates about post-World War I America. An examination of the reactions of the miners' union, the coal industry, and the state and national governments to the chain of events set in motion by the Massacre fosters a new understanding of the relationship between these groups.[41]

In the aftermath of the First World War, the United States' largest labor union squared off against the country's most powerful corporation. Beginning in 1919, the United Mine Workers of America and the United States Steel Corporation fought repeatedly in a monumental struggle over the direction of the nation's industrial development. In the decades that followed, the struggle between these two giants transformed American labor relations, the role of government in the economy, and the meaning of individual rights in a modern society.[42]

One of the first battlegrounds was West Virginia, where the armies that took the field were motivated by philosophies forged in an era that had already passed. The majority of District 17's officers were West Virginians who believed strongly in the customary rights that came from an elemental connection to their native soil. Their desire to promote the collectivization of their fellow miners reflected a devotion to protecting the inherent dignity of men who toiled in the bowels of the earth. Their opponents in the Williamson-Thacker strike of 1920–1922 believed just as strongly in the absolute right of property. Although these equally intractable groups fought for two and half years over the unionization of southern West Virginia, neither side gained from the struggle; instead, the benefits accrued to their external allies. To understand how both John L. Lewis and the corporate giants of southern West Virginia's coal industry profited from the bloody and protracted strike, one must first understand how the local combatants wound up fighting a war of attrition that allowed their allies to move in, take over, and transform the conflict.[43]

When the organizers of District 17 arrived in Mingo County in the spring of 1920, they found a uniquely receptive audience. Unlike their compatriots in Logan and McDowell, almost half of Mingo's miners were native, white, and the descendants of the Tug Valley's original settlers. When organizers preached about the need to unionize in order to defend against the encroachment of "capitalists from New York and London," Mingo's miners understood the message. In turn, the leaders of the national and district leadership of the UMWA considered Mingo County an ideal beachhead for their renewed effort in southern West Virginia. First, in the past, Mingo's miners had responded repeatedly to the union's call. Second, the sheriff of the county had defied the coal operators and refused to intervene against the union. Third, the venality of local politics ensured that at least one faction would ally with the union in the hopes of gaining primacy over the county. Fourth, should the situation turn dangerous, the proximity of the Kentucky border offered an escape route for beleaguered organizers. Fifth, the Williamson-Thacker coalfield was a peripheral concern; the operators were focused instead on keeping the union out of Logan and McDowell, the largest coal producing counties in the state. Last and most important, if or when violence broke out, the union could disclaim responsibility because after all, Mingo was Hatfield-McCoy country, and the county's reputation spoke for itself.[44]

Despite all of these advantages, District 17 squandered its early organizational success in Mingo by committing a series of strategic blunders. Internal divisions among District 17 leadership further undermined the Williamson-Thacker drive. Because, like the operators, the union focused on Logan and McDowell Counties, District 17 leaders failed to understand the goals of Mingo's miners and the work conditions that prevailed in the Williamson-Thacker coalfield. The operators seized each of these mistakes, and as the strike wore on used them to undermine the miners' resolve and the union's public integrity.[45]

In the beginning, the primary demand of Mingo's miners had been for the 27-percent wage increase which had been granted to union miners at the end of March 1920. District 17 Secretary Fred Mooney knew that the miners in southern West Virginia's non-union coalfields did not qualify for the increase and had expressed displeasure at what he viewed as Bill Blizzard's demagogic manipulation of miner discontent over the issue. Still, the union organizers did nothing to dissuade the miners' expectation of just such a reward. The Williamson Operators' Association struck an early blow to the union's credibility by making the wage increase the subject of its first published propaganda. "That 27 Percent Increase" informed Mingo's miners that because the Bituminous Coal Commission awarded the increase only to union miners bound by the terms of the Washington Agreement, the UMWA could not compel the federal government to grant it to miners organized after the war. Translation: under no circumstances were Mingo's miners eligible for the increase, and their employers could not be compelled to grant it. Eventually, as the lockout-turned-strike dragged on and Mingo's miners faced increasing numbers of replacement workers who were enjoying the benefits of inflated wages, the union lost its credibility and the support of many miners.[46]

The union coupled its disingenuous manipulation of the wage issue with an appalling ignorance of work conditions in Mingo County. Despite having declared Mingo completely organized, and in spite of responding to the operators' lockout with a strike call, the union failed to shut down coal production in the county for more than a few days. One of the largest companies, Borderland, never ceased production. Within weeks, the only organized mines in Mingo were the wagon mines and four modest-sized tipple mines. As late as July 1921, District 17 President Frank Keeney obstinately refused to acknowledge the full import of this turn of events. In a letter to Governor Ephraim F. Morgan, which he later read before the Senate investigating committee, Keeney outlined the union's six demands for resolution of the Williamson-Thacker strike. The second and third demands called for semi-monthly paydays and an eight-hour work day. When Governor Morgan inquired of Mingo's acting prosecuting attorney, S. D. Stokes, about Keeney's demands, Stokes pointed out to the governor that the Williamson operators' association had voluntarily instituted semi-monthly pay in 1913 and the eight-hour day in 1919. The union had overlooked the fact that because wagon mines did not qualify for membership in the operators' association, they were the only mines that had not already granted those concessions. Unfortunately for Keeney, his apparent ignorance of the situation was gleefully pointed out before the Senate committee by operator attorney S. B. Avis.[47]

As the operators whittled away at the union effort in Mingo, District 17's internal divisions emerged and further weakened the union's position. Never a united group, several of the officers of District 17 disliked and distrusted each other even before the beginning of the Mingo organization drive. District 17 internal politics and the distribution of strike-related funds further complicated the situation. Former district officers accused Frank Keeney of letting Mingo's children go hungry in order to spend money getting re-elected district president in December 1920. By 1921 coal operators were gleefully discussing rumors that District 17 officers were barring an outraged Mother Jones access to "the books." The two issues that further complicated relations between the district officers and the organizers sent to Mingo were politics and the role of violence in the struggle. As in their misreading of the labor situation, District 17's lack of understanding regarding local politics handed the operators another advantage. The union's contradictory stance on the violence committed by Mingo's miners also served the operators' agenda.[48]

Although the union organizers who first entered Mingo stumped for Republican Samuel Montgomery, the most successful local alliance the union had forged was with the Democratic leadership of the county. Angered by the union's early alliance with the Democrats and then the splinter movement led by District officers in support of Montgomery and the Non-Partisan League, Mingo's Republicans schemed to weaken the union's political pull in the county. The union's ties to Mingo's Democrats and then the Non-Partisan League also undermined its appeal to the African-American population, both within the county and in the surrounding coalfields. The operators lured away Judge Damron, whose kinsman, Massacre-participant Isaac Brewer, later became a star witness against his fellow defendants.[49]

Perhaps the most damaging miscalculation made by District 17 in the strike was its mishandling of the violence that frequently erupted in Mingo County between 1920 and 1922. As the conflict dragged on, Mingo Countians grew tired of living in a war zone. West Virginians resented the drain on public funds and the notoriety the violence brought the state. Journalists, while primarily intending to elicit support for the union, sensationalized tales of murder and mayhem and ultimately diminished national support for the miners' cause. Union funds were exhausted by the legal costs of defending striking miners' attacks on replacement workers, law officers, and private citizens. Privately, union attorneys and organizers expressed distaste for the miners' actions. In fact, one organizer, A. E. Hester who left the county in disgust only to be accused of embezzling strike funds ultimately testified against the union. Even UMWA President John L. Lewis steered clear of West Virginia, returning to the state only when the officers of District 17 were tried for treason following the Battle of Blair Mountain. In turn, the operators used each incident to justify escalating state and federal intervention, supposedly intending to restore "law and order" in Mingo County but in reality hoping to break the strike.[50]

The cumulative effect of the violence can be measured by the conflicting outcomes of the Senate investigations in 1913 and 1921. The 1913 investigation into the Paint Creek and Cabin Creek strike of 1912–1913 had helped force a resolution of the strike which, at least in some ways, benefited the affected miners. By contrast, despite the involvement of at least one U.S. senator from the 1913 investigation, the 1921 committee failed to even file a report of its findings. District 17 officers, who had agitated for a senate investigation as early as June 1920, watched helplessly as their own leaders transformed the hearings from an investigation into conditions in Mingo and anti-union southern West Virginia into a public denunciation of U.S. Steel.[51]

After spending nearly eight million dollars on the West Virginia unionization drive of 1920–1922, the union admitted defeat in October 1922. Although District 17's officers escaped legal punishment for strike-related activities in Mingo and charges of treason following the Battle of Blair Mountain, their days as leaders of West Virginia's miners were numbered. John L. Lewis, who had barely won reelection as UMWA president in 1922, forced Keeney and Mooney to resign in 1924. Lewis simultaneously stripped District 17 of the right to autonomously elect its officers. Until his own death in 1960, Lewis maintained his grip on West Virginia's miners; from District 17 headquarters down to the local level, union officers held their positions based on their loyalty to Lewis. Frank Keeney, Fred Mooney, and Bill Blizzard all died broken men, outcasts from their union. In contrast, John L. Lewis emerged a hero from the chaos of the 1920–1922 strike and the union's near disintegration in the 1920s and early 1930s.[52]

Like the leadership of District 17, the operators of the Williamson-Thacker coalfield were also unwitting foot soldiers of a larger army. Their failure to contain the 1920 unionization drive resulted in Mingo's capitulation to the interests of neighboring competitors and ultimately, the absorption of the Williamson-Thacker coalfield into the U.S. Steel coal empire. A brief overview of the strike from the operators' perspective reveals how the corporate giants of southern West Virginia's coal industry used the strike, not only to defeat the union but also to drive out their smaller competitors.[53]

Dismissed by other southern West Virginia operators as "gentry" who had to be "forced" into adopting anti-union policies, the Williamson-

The Lick Creek Tent Colony, ca. 1922. *The Lick Creek tent colony challenged the established order in Mingo County on several levels. [Library of Congress Prints and Photographs Division; Negative number LC-F82-7373]*

Thacker operators made several critical mistakes that ultimately cost them autonomy. First, they failed to enforce their "no mediation" stance throughout the coalfield. Although Red Jacket, Borderland, and Stone Mountain moved quickly to evict union miners, other companies waited until federal troops had come to Mingo for the second time in December 1920. Second, even after the might of the state and federal governments had been brought to bear on the strike, the coal and political leadership of the county allowed themselves to be "led astray on side issues" that prolonged the effort to restore law and order. A dispute over whom Governor Morgan would recognize as the commander of the Mingo Militia illustrates that not even the strike could prevent the habitual squabbling of the county's elites. Third, the operators resorted to tactics that the union could expose and manipulate in the court of public opinion; for example, several operators faced charges of bribery and complicity in agent provocateur activity.[54]

The most egregious and ill-timed act by the operators was the raid on the Lick Creek tent colony on June 14, 1921, the same day that the West

Virginia State Supreme Court of Appeals ordered the release of three men who had been jailed for infractions against an illegal enacting of martial law. Since June 14 was also Flag Day, the attack seemed to embody the struggle over which side was fighting in defense of "American" values, the union or the force of state police and local vigilantes who trampled the colony's flag and killed one miner. Not long after the Lick Creek incident, the United States Senate finally approved an investigation into conditions in West Virginia. Distracted by the chaos that had swept over the county, the Williamson operators were forced to seek alliances with their more powerful neighbors.[55]

The extent to which Mingo's elites bungled their handling of the strike is most evident when their actions are contrasted with the reaction of operators in the neighboring coalfields. During the early heady days of the union drive in Mingo, the operators of the Guyan, Pocahontas, and Winding Gulf coalfields quickly moved to erect a *cordon sanitaire* around Mingo. Wage hikes were ordered and compulsory donations to the operators' "insurance" fund were raised. To better coordinate their activities, the Tug River and Winding Gulf operators' associations moved their headquarters to Welch, where the Pocahontas operators were based and where a regional office of the Baldwin-Felts agency had been established. When strike-related disturbances spread to the surrounding coalfields, the operators again acted vigorously. In Logan County, Sheriff Chafin received the necessary funds to enlarge his army of deputies. When coal company property on the Mingo-McDowell border was attacked, McDowell Sheriff S. A. Daniel requested state police officers from Governor Cornwell. The most decisive if ill-advised and reprehensible action that was supported, if not orchestrated, by the operators was the retaliatory murder of Sid Hatfield and Ed Chambers on August 1, 1921.[56]

Even while the operators of the surrounding coalfields worked to contain the Williamson-Thacker strike, they also scrambled to turn the Mingo operators' trouble to their own benefit. Union miners from McDowell County, West Virginia, and Pike County, Kentucky, were driven into Mingo's tent colonies, exponentially raising the union's relief costs. When the UMWA launched the Mingo strike, the N&W coal cars assigned to the Williamson-Thacker district were diverted to the Tug

River and Pocahontas coalfields. Because the strike also coincided with a resurgence in the demand for coal while the mines in Mingo were idle or producing at diminished capacity, Tug River and Pocahontas operators commanded $9 a ton. The *coup de grace* for many Mingo operators came when their mines were gobbled up by giant corporate interests. In 1920, for example, Williamson entrepreneur E. L. Bailey sold out to the Solvay combine, a subsidiary of the American Rolling Mills Company (however, he stayed on as superintendent). Later that same year, the Thacker Fuel Company purchased the Grey Eagle Coal Company of Kermit. Thus, the Williamson-Thacker strike's failure had not resulted from Mingo County's membership in the anti-union southern West Virginia monolith; instead, the strike gave the monolith the much needed opportunity to "quietly" but "effectively" move in and defend its Achilles' heel.[57]

The reaction of the state and federal governments to the 1920–1922 West Virginia conflict also illuminates much about post-World War I America. In fact, the second West Virginia mine war could hardly have been more ill-timed. Wartime regulations had demonstrated that the bureaucratization of American society, and more important, the American economy, could result in efficient production expansion. Politicians and business leaders emerged from the war aware that a new era of business and government relations was dawning. Convinced more than ever that elites should be entrusted with directing the ship of state, politicians and business leaders set about engineering the consent of the society they governed. As a result, the national debate on civil liberties was transformed. Voices of dissent, especially those that rose from the working class, were denounced and stifled. After decades of upheaval and reform, the American public fell in behind the country's business and political leaders and entered the 1920s intolerant of challenges to the status quo. In 1920, the average American was more interested in the maintenance of a steady supply of coal than in the individual human rights of West Virginia's miners.[58]

Against this backdrop, the political and business elite of West Virginia declared themselves the guardians of the nation's most important fuel supply and proceeded to wage war on the rights of the state's working class. When a local jury acquitted Sid Hatfield and the other Massacre

236

defendants, the state legislature passed a bill that abrogated defendants' rights. Although the West Virginia State Supreme Court of Appeals later criticized abuse of the law, it did not overturn it. A former West Virginia circuit court judge even advocated the abolition of juries because, in his opinion, they all too often acted like mobs. The West Virginia legislature also passed a law, proposed by Mingo's state senator M. Z. White, that empowered coal companies to seek compensation from unions for property damage suffered during strikes. Inspired by the example of the injunction cases used to kill the Williamson-Thacker strike, West Virginia's coal operators spent the 1920s slowly strangling and bleeding the UMWA dry. By 1928, over 200 injunction cases had been won by the state's coal elite; miners were prevented from parading on public highways and meeting on private property or even in churches. Denied the basic constitutional right of freedom of speech, West Virginia's miners believed that they not only had been silenced, but enslaved.[59]

The federal government's collusion in West Virginia's mistreatment of the miners stemmed primarily from inertia. While generally supportive of the rights of industrial workers, national leaders on both ends of the political spectrum perceived themselves primarily as the protectors of the public interest. As a result, the federal government became an often unwilling collaborator in the oppression of West Virginia's miners. Unable to compel the anti-union operators to negotiate a settlement of the Williamson-Thacker strike, Washington could only step in and restore peace when the struggle lapsed into armed confrontation.[60]

The United States Supreme Court provided little guidance in the resolution of labor conflicts because for decades it simultaneously recognized a small number of union rights while primarily protecting the interests of the open shop. Only violent public upheaval seemed to motivate federal legislators into action, which generally consisted of impaneling investigative committees or commissions. By the time of the Williamson-Thacker strike, the efficacy of the seemingly endless string of senate committees and federal commissions was openly questioned. Known as "that puttering and futile investigating committee," the 1921 committee entrusted with the examination of conditions in West Virginia not only failed to arrive at a solution for the state's labor woes, its chairman also managed to antagonize both sides in the controversy. The inability or unwillingness of the federal government to guide the course of labor relations in the 1920s contributed to the arrival of "desolate days in America."[61]

While the Matewan Massacre resulted from a unique convergence of local tensions, it also stands as a tragic example of the inherent conflicts that have periodically erupted throughout our nation's history. In the spring of 1920, Mingo County steamed like a pressure-cooker under the accumulated weight of decades of political corruption and chronic economic instability. When violence swept through the streets of Matewan, the local community, West Virginia, and America were all forced to confront the issues that had precipitated the fatal confrontation. Unable to arrive at a consensus over the meaning of property versus individual rights, the residents of Mingo County divided and perpetuated a cycle of mutual exploitation and denigration. After the bloody rampage, the survival of any semblance of community necessitated a retreat behind a wall of silence. As one Matewan native observed, "How do you explain to a child that the nice old man who runs the post office was once indicted for murder?" It is not uncommon for current residents to count as an ancestor someone from both sides of the conflict.[62]

Beyond the boundaries of Mingo County, the meaning of the Matewan Massacre was surrendered to the agendas of whoever invoked it. The United Mine Workers of America transformed the ambush of the Baldwin-Felts agents into a righteous attack on the forces of oppression, by which they hoped to inspire all of West Virginia's miners to rise. To the proponents of the open shop and welfare capitalism, the Massacre demonstrated the disruptive and destructive influence of trade unionism that must be defeated in order for liberty to survive. Liberals and conservatives alike manipulated details of the story in order to frighten Americans into supporting industrial reform. However, the Massacre and the two years of strife it ushered in evaded ultimate definition and thus proved of limited utility to any one faction in the body politic. The American people were already accustomed to dismissing Appalachian violence as the outbursts of a backward subculture. Because both the defenders and the critics of the events in Matewan on May 19, 1920, felt compelled to note that the little town on the Tug was also the home of the "feudin' Hatfields and McCoys," the Massacre passed into the realm of myth. Secure in their own modernity and prosperity, the rest of the United States ignored the ravages of the war being waged between the forces of capitalism and unionism. Until the dark days of the Great Depression when the wail of another Appalachian community torn asunder echoed through the hills of Eastern Kentucky, Americans turned a blind eye and a deaf ear.[63]

The contrast between what was born of the suffering in "Bloody Mingo" and "Bloody Harlan" illustrates the callousness of contemporary American society. In a nation weary of reform and intoxicated by the wealth of the post-World War I economy, the miners of Mingo and southern West Virginia were acceptable casualties; after all, had these people not always been unfortunate reminders of our violent past? At one point during the struggle, the *New York Times* had editorialized that the union and the miners should be forced to abandon their unhealthy tent colonies for centrally located camps that could be watched and protected. By contrast, just one decade later, to a nation brought to its knees by utter economic collapse, the miners of eastern Kentucky became a symbol of America's dire need of change. The head of yet another federal commission, liberal crusader Robert La Follette sent Heber Blankenhorn, who had also documented events in West Virginia, into Harlan. Armed with the information gathered by Blankenhorn, the federal government indicted the coal operators and their deputies for their crimes. In search of votes, Franklin Roosevelt and the Democratic Party courted organized labor. Once in office, when Roosevelt and his congressional allies set about repaying their debt to the unions, they enlisted the aid of Jett Lauck, who in his 1921 Senate testimony had compared West Virginia's anti-union operators to slaveholders. Lauck would help draft the National Recovery Act.[64]

But what of Matewan, Mingo County, and the miners of southern West Virginia? The news of the passage of the National Industrial Recovery Act sent the United Mine Workers of America sweeping back into the coalfields. Inspired by the union's revival, the former "mayor" of the Lick Creek tent colony joyously observed, "All the demons in Hell can't keep us from organizing ... now!" However, in the decades that followed, members of the generation traumatized by the events of 1920–1922 often fell victim to divorce, alcoholism, madness, and "the ugliness of their memories." For others, to forget was to heal. Because some healing did take place, the son of a Massacre participant preached the funeral of a woman driven from her home because of what she had seen that day. The 1987 release of John Sayles' film *Matewan* renewed interest in the story and facilitated funding for an oral history initiative. The rediscovery of the Matewan Massacre could have meant the reopening of old wounds. Instead, it became a path for a community in search of its own past and a way for the people of Matewan to reclaim their story.[65]

NOTES

1   *Williamson Daily News*, May 20, 1920.

2   *Williamson Daily News*, January 13, 1920, March 23, 1920, March 25, 1920.

3   Barkey, 214; *WVCF: Personal Views*, 20; Mildred Allen Beik, *The Miners of Windber: The Struggles of New Immigrants for Unionization, 1890s–1930* (University Park, PA: The Pennsylvania University Press, 1997 [1996]), 260.

4   Bittner's testimony was reported in *The Coal Trade Journal* 51 (February 4, 1920): 13–14; "John L. Lewis, Head of United Mine Workers of America, Here for Purpose of Organizing Local Coal Fields," *Bluefield Daily Telegraph,* January 31, 1920.

5   *Coal Trade Journal* 51 (January 21, 1920): 25; *Coal Trade Journal* 51 (January 7, 1920): 13, and 51 (February 4, 1920): 120; E. F. Striplin, *The Norfolk and Western: A History* (Roanoke, VA: Norfolk and Western Railway Company, 1981), 153; Federal regulation ended on March 1, 1920; *Coal Trade Journal* 51 (February 18, 1920): 174; "Testimony of W. E. Hutchinson," *West Virginia Coal Fields*, 79.

6   City Attorney S. D. Stokes to Hunt-Forbes Coal Company, January 8, 1920, Stokes Papers, WVRHC; Court Documents relating to the case, in Box 5 of the Stokes Papers, WVRHC.

7   *Coal Trade Journal* 51 (January 7, 1920): 17; *Coal Trade Journal* 51 (March 17, 1920): 292, and 51 (March 24, 1920): 319.

8   "Flurry Caused By Invasion of Head of Mine Workers," *Bluefield Daily Telegraph*, February 1, 1920; *Bluefield Daily Telegraph*, February 7, 1920.

9   Merle T. Cole, "Martial Law and Major Davis as 'Emperor of Tug River,'" *West Virginia History* 43 (Winter 1982): 118–144, 126.

10   *Williamson Daily News*, March 30, 1920; *Williamson Daily News*, April 1, 1920; *Coal Trade Journal* 51 (April 7, 1920): 368.

11   "Testimony of Frank Ingham," *West Virginia Coal Fields*, 29.

12   "Testimony of W. E. Hutchinson," *West Virginia Coal Fields*, 80–81. Who placed the notice at the mine entrance remains a mystery.

13   "Testimony of W. E. Hutchinson," *West Virginia Coal Fields*, 83–85; *Williamson Daily News*, April 23, 1920.

14   *Williamson Daily News*, April 27, 1920; *Williamson Daily News*, April 29, 1920.

15   *Williamson Daily News*, April 29, 1920; *Charleston Gazette*, May 20, 1920; *Charleston Gazette*, May 20, 1920; "Testimony of C. [sic] T. Blankenship," *West Virginia Coal Fields*, 490. During the testimony of Harry Olmsted, coal company attorney S. B. Avis noted that prior to the beginning of the strike on July 1, 1920, only 20 evictions: a total of 16 from two separate occasions at Matewan, 2 at Red Jacket, and 2 at Burnwell had taken place; by the time the Senate Hearings began in July 1921, a total of 369 evictions had been processed in Mingo County. Avowal of S. B. Avis, during "Testimony of Harry Olmsted," *West Virginia Coal Fields*: 223–271, 257.

16   "Testimony of W. E. Hutchinson," *West Virginia Coal Fields*, 92; Avowal of S. B. Avis, during "Testimony of Harry Olmsted," *West Virginia Coal Fields*: 223–271, 257; "Testimony of T. L. Felts," *West Virginia Coal Fields*, 881–905, 891; Warner, 373; Source for Stone Mountain contract: *United Mine Workers Journal* 51 (February 15, 1921):17; Source for Red Jacket contract: "Affidavit of C. L. McKinnon," *Red Jacket, et al v. John L. Lewis, et al*: 788a–791a, 789a. McKinnon testified that Red Jacket imposed its contract on May 10, 1920, following a union meeting at Matewan.

17   *Charleston Gazette*, May 20, 1920; Daniel P. Jordan, "The Mingo War: Labor Violence in the Southern West Virginia Coal Fields, 1919–1922," *Essays in Southern Labor History: Selected Papers, Southern Labor History Conference, 1976*, edited by Gary M. Fink and Merl E. Reed, *Contributions in Economics and Economic History*, no.16 (Westport, CT: Greenwood Press, 1977), 107; *Williamson Daily News*, May 15, 1920; Report of Operative #24, dated May 14, 1920, submitted to George Bausewine by T. L. Felts, May 17, 1920, reprinted as part of "Hatfield Exhibit No. 2," in "Testimony of Sid Hatfield," *West Virginia Coal Fields*: 205–221: 215; *Bluefield Daily Telegraph*, May 15, 1920.

18   *Williamson Daily News*, May 8, 1920; *Williamson Daily News*, May 14, 1920.

19   Shifflett, 123; Arthur Gleason, "Public Ownership of Private Officials," *Nation* 110 (May 29 1920): 724–725, 724.

20   "Testimony of W. E. Hutchinson," *West Virginia Coal Fields*, 80; *Charleston Gazette*, May 20, 1920; "Testimony of Sid Hatfield," *West Virginia Coal Fields*, 219; See: Billings and Blee, *The Road to Poverty: The Making of Wealth and Hardship in Appalachia.*

21   Source for Johnson: Wesley M. Bagby, *The Road to Normalcy: the Presidential Campaign and Election of 1920* (Baltimore: Johns Hopkins Press, 1968 [1962]), 31–32; Source for Kenyon: *Bluefield Daily Telegraph*, February 1, 1922; Robert K. Murray, *Red Scare: A Study of National Hysteria, 1919–1920* (New York: McGraw Hill, 1964), 30–31, 206; Source for Wood's appearance: *Williamson Daily News*, May 20, 1920 (nothing has been found to indicate that Wood did not appear); *Charleston Gazette*, May 21, 1920; See member list of the Committee that follows the title page, *West Virginia Coal Fields*, (1921).

22   *West Virginia Heritage Encyclopedia*, volume 15: 3302; Evelyn L. K. Harris and Frank J. Krebs, *From Humble Beginnings: West Virginia State Federation of Labor, 1903–1957* (Charleston, WV: Jones Printing Co., 1960), 119; *Williamson Daily News*, May 15, 1920; Source for Morgan: *West Virginia Heritage Encyclopedia*, vol. 15: 3340–3341; Source for Grosscup: *West Virginia Heritage Encyclopedia*, vol. 10: 2064–2065.

23   The *Williamson Daily News*, a Democratic paper described the months-long machinations of the three Republicans gubernatorial hopefuls in its May 15, 1920 issue; *Bluefield Daily Telegraph*, January 18, 1920; A. D. Sowers, *Some Facts about McDowell County, West Virginia* (Keystone, WV: A. D. Sowers, 1912), in Rare Book Collection, WVRHC.

24   *Williamson Daily News*, March 9, 1920.

25   *Williamson Daily News*, May 15, 1920.

26   *Williamson Daily News*, May 15, 1920.

27   S. D. Stokes to Carl E. Whitney, August 19, 1920, Stokes Papers, WVRHC.

28   *Bluefield Daily Telegraph*, January 18, 1920; *Williamson Daily News*, May 15, 1920.

29   "Testimony of C.E. Lively," February 25–26, 1921, unknown newspaper, Matewan Omnibus Collection, ERCA; "State ex rel. M. Z. White, Relator v. County

Court of Mingo County," *Reports of the West Virginia Supreme Court of Appeals* 86 (March 16–September 21 1920): 517–518.

30 *Williamson Daily News*, January 30, 1920; See Tudiver, 113 and Billings and Blee, 281; *Williamson Daily News*, May 25, 1920.

31 *Williamson Daily News*, May 25, 1920; Spivak, *A Man in His Time*, 87; Houston G. Young, Secretary of State of West Virginia, *State of West Virginia: Official Returns of the General Election, held November 2, 1920* (Charleston, WV: Tribune Printing Co, 1920), J. Hop Woods Bound Pamphlet Collection, 11476, volume 15; Hatfield correspondence, no.16.

32 Despite what has been written for decades about the exclusion of miners from the exercise of power locally, before the Massacre miners often held positions of power in the community; for example, miner Elihu Boggs served both as mayor of Matewan and clerk of the county court *Mingo Republican*, January 2, 1914; Source for Boggs as clerk of county court: *Williamson Daily News*, November 26, 1920; Massacre defendant Fred Burgraff had served as deputy sheriff: *Williamson Daily News*, June 2, 1916. See Eller, 10; See also Herbert Gutman, "Joseph P. McConnell and the Workers' Struggle in Paterson, New Jersey," in Herbert Gutman, *Power and Culture: Essays on the American Working Class*, edited by Ira Berlin (New York: Pantheon Books, 1987), 93–116; Bertha Damron, interview with Rebecca J. Bailey, Summer 1989 Matewan Oral History Project; Source for violence in defense of community, especially "women and children": See Stock, *Rural Radicals: Righteous Rage in the American Grain*; Source for Appalachian criminals turned folk heroes: Montell, 138; Source for Sid Hatfield as miners' hero appeared during the Massacre trial. Mrs. Stella Scales, who had been called to testify about the statements and actions of prosecution witness Joe C. Jack, was asked if she also knew Sid Hatfield. After responding that she did not, she added that she would like to and also wished to "shake his hand." "Testimony of Mrs. Stella Scales," Lewis Collection, ERCA.

33 Coleman, 100; these same organizations did extend aid to miners' families in Fayette and Raleigh Counties, *Bluefield Daily Telegraph*, January 27, 1922; Source for "proposition": "Mine Dynamited On Matewan Day" *New York Times*, May 20, 1921.

34 Hatfield correspondence, letters no. 16 and 24; T. E. Bowman, "From West Riding"; Tudiver, "Political Economy and Culture in Central Appalachia," 122; "Statement of Mrs. Billy Duty," report by #19, September 9, 1920, Lewis Collection, ERCA.

35   Jeannette Simpkins, interview with Rebecca J. Bailey, Summer 1990 Matewan Oral History Project; "Testimony of J. R. Brockus," *West Virginia Coalfields*, 344; Edward L. Ayers, *The Promise of the New South: Life After Reconstruction* (New York: Oxford University Press, 1992), 408; Jacquelyn Dowd Hall, et al, *Like a Family: The Making of a Southern Cotton Mill World* (Chapel Hill and London: The University of North Carolina Press, 1987), 126, 178–179; Virginia Grimmett, interview with Rebecca J. Bailey, Summer 1989 Matewan Oral History Project; "From Nolan, W.Va,"*United Mine Workers Journal* 32 (April 15, 1921): 15. Identified only as the wife of Borderland #2 local's secretary, the correspondent wrote to the *Journal*, "I am a union woman. . . . We are more de-termined than ever to win. We are like the Holiness song, 'we are determined to hold out to the end'"; Hatfield correspondence, letter no. 16.

36   *Williamson Daily News*, November 4, 1920; *Cincinnati Enquirer*, May 28, 1921, Matewan Omnibus Collection, ERCA; "Report of # 9," July 29, 1920, Lewis Collection, ERCA; "Record of Mrs. Sid Hatfield," Lewis Collection, ERCA; *Williamson Daily News*, December 13, 1920; "Statement of Mrs. Pearl Hatfield," November 1, 1921, "Berman Hatfield Case," Lewis Collection, ERCA; The use of effigies and mock funerals can be traced to England, For more see: E. P. Thompson, *Customs in Common* (New York: New Press: Distributed by W. W. Norton, 1991), 480; Bertha Damron interview; the allegations made by Mrs. Pearl Hatfield against John Collins in 1921 were unknown at the time of the interview, and thus were not discussed; *Williamson Daily News*, August 16, 1920. Among those who left Matewan were: Charlie McCoy, the town barber, George Gunnoe, the principal of Matewan High School, Dr. Wade Hill, and John and Mary Brown, the proprietors of Matewan's laundry. Source for McCoy departure: Hatfield cor-respondence, letter no. 5; Source for Gunnoe departure: "Affidavit of George H. Gunnoe," Lewis Collection, ERCA; Source for Hill departure: Lee, *Bloodletting*, 61; Source for Browns' departure: Johnny Fullen interview. The term "exodus" is a direct quote from "Good People Flee Matewan," *New York Times*, 21 May 1921; Dixie Accord interview; Rufus Starr interview; "Testimony of A. E. Hester," *West Virginia Coal Fields*, 826; E. K. Beckner to John J. Cornwell, July 13, 1920, Cornwell Papers, WVRHC; Jim Backus, interview with John Hennen, Summer 1989 Matewan Oral History Project; The principal recipients of the miners' gratitude were the Chambers' and Beckner & Hynes. Dixie Accord interview; Jim Backus interview; Johnny Fullen interview; (Joseph and Samuel Schaeffer) Abraham J.

Shinedling, *West Virginia Jewry: Origins and History* 3 vols. (Philadelphia: Maurice Jacobs, Inc, 1963), volume 2: 1038; "Statement of Eli Sohn" *State of West Virginia v. C.E. Lively, George Pence, and William Salter,* Lewis Collection, ERCA.

37   Union organizers sent into Mingo also sought to secure the miners' support for the candidacy of Samuel Montgomery in the 1920 Republican gubernatorial primary. Various sources allude to the political undercurrents that swirled in Matewan at the time of the Massacre. See *Williamson Daily News,* February 20, 1919; "Sid Hatfield by Isaac Brewer," Lewis Collection, ERCA; "Trial Testimony of Elizabeth Burgraff," Lewis Collection, ERCA; "Trial Testimony of Hugh Combs," Lewis Collection, ERCA. For more on the Sid Hatfield-A. B. Hatfield incident see author's dissertation chapter nine; Broadside "Remember Your Friends—Also Your Enemies," Lewis Collection, ERCA. This pro-union handbill, listed who had contributed to the bonds posted by the men on both sides who had been indicted as a result of events surrounding the Massacre and strike-related activities; "Report of #9," August 6, 1920, Lewis Collection, ERCA. Undercover agent #19 reported that he had heard an International (UMWA) officer claim that he had spoken with Damron, who promised to do all he could for the Massacre defendants. "Report of #19," August 5, 1920, Lewis Collection, ERCA; *Coal Trade Journal* 52 (January 5, 1921): 7. The Bishop-Hatfield mine may have been a wagon mine and, as such, did not appear in the annual reports of the West Virginia Department of Mines.

38   *Williamson Daily News,* October 12, 1920; *Bluefield Daily Telegraph,* January 19, 1921. Damron's betrayal was not forgotten. Years later he was shot by an unknown assailant. Lee, *Bloodletting,* 51; Pinson offered no resistance to the imposition of martial law. See correspondence between Governor Cornwell and various Mingo County elites regarding Blankenship from late November 1920. Cornwell Papers, WVRHC; Richard Burgett interview, *On Dark and Bloody Ground: An Oral History of the U.M.W.A. in Central Appalachia, 1920–1935,* edited by Anne Lawrence (Charleston, WV: Miner's Voice, 1973): 105–109, 107; *West Virginia Coalfields,* 486.

39   *West Virginia Bluebooks,* 1925–1935; Source for Blankenship never re-entered politics: Tom Blankenship interview with C. Paul McAlister, Summer 1990 Matewan Oral History Project; Source for Thurman "Broggs" Chambers supplied the miners with guns: Rufus Starr interview; Source for accusation regarding Dan Chamber: S. B. Avis avowal, "Testimony of C. E. Lively," *West Virginia Coal*

Fields, 355; Hatfield correspondence, letter no. 16. Among the four Mingo political bosses indicted in 1970 for subverting the 1968 general election, two, Harry Artis and Arnold Starr were from families prominent in the 1920–1922 strike. Perry, *They'll Cut Off Your Project*, 252; Source for Perry as a boss: "James Washington interview," in *On Dark and Bloody Ground*, 119–121, 119. The political loyalty of Mingo's miners can be interpreted as an example of the American Federation of Labor's policy of "reward your friends, and punish your enemies." Quote source: Laslett, *Colliers*, 231; See: Perry, *They'll Cut Off Your Project*, 155–156, 209–210 and 255; A. D. Lavinder interview, WVRHC; Venchie Morrell (1990) interview.

40   Jim Backus interview; Mooney, *Struggle in the Coalfields*, 127–128; See also Chapter 10 "Catastrophe in Coal" of Bernstein's *The Lean Years*. Irving Bernstein, *The Lean Years: A History of the American Worker, 1920–1933* (New York: Da Capo Press, 1983 [1960]), 358–390; Edmund Wilson, "Frank Keeney's Coal Diggers," *New Republic* 67 (July 8, 1931/15 July 15, 1931): 195–199, 229–231; and Hennen, *Americanization of West Virginia*, 105; Conley, *History of the West Virginia Coal Industry*, 262; Source for "riches": Johnson, *Politics of Soft Coal*, 95.

41   E. P. Thompson, *The Making of the English Working Class* [New York: Vintage Books, 1966], 9). Anatomise means to reduce to discrete causative factors.

42   Source for UMWA: Bernstein, 127; Source for U.S. Steel: Chernow, 82.

43   David A. Corbin, "'Frank Keeney is Our Leader and We Shall Not Be Moved': Rank and File Leadership in the West Virginia Coal Fields," in *Essays in Southern Labor History: Selected Papers from the Southern Labor History Conference, 1976* edited by Gary M. Fink and Merle E. Reed, *Contributions in Economics and Economic History*, no. 16, (Westport, CT: Greenwood Press, 1976), 147; Mooney, *Struggle in the Coal Fields*, 13, 16; "Testimony of W. Jett Lauck," *West Virginia Coal Fields*: 1045; Source for Vinson and the operators' association: McIntosh Memoir, WVRHC; Source for Vinson on UMWA: "Opening Statement of Z. T. Vinson," *West Virginia Coal Fields*, 8–15; "War of attrition" was borrowed from the memoir of journalist Jack Spivak, a participant-observer of the Williamson-Thacker strike. Spivak, *A Man in His Time*, 102.

44   45 percent of the men who worked in Mingo's mines in 1920 came from families that had lived in the greater Big Sandy Valley since before 1850; Corbin, "Frank Keeney Is Our Leader," 147; See author's dissertation chapter four; *United Mine Workers Journal* 11 (July 4, 1901): 2. See author's dissertation chapter four;

Mingo Republican, June 3 1920; "Seven Prisoner After Mingo Battle," *New York Times,* May 27, 1921; See Lunt, 147, 152–153; "Testimony of C. F. (Frank) Keeney," *West Virginia Coal Fields,* 184.

45 *New York Globe,* August 27, 1920; Gleason, "Private Ownership," 724.

46 Testimony of W. E. Hutchinson," *West Virginia Coal Fields,* 74; *Williamson Daily News,* March 30, 1920; Mooney, *Struggle in the Coal Fields,* 67; "That Twenty-Seven Per Cent Increase," *Williamson Daily News,* June 2, 1920; *Williamson Daily News,* October 19, 1920. The *News* reported that for the first time since "the war" there were more men than jobs in Mingo.

47 Source for Lockout: *Coal Trade Journal* 51 (June 5, 1920): 22; Source for strike: "Testimony of C. F. Keeney," *West Virginia Coal Fields,* 104; District 17 sent strike notices to 70 mines, miners at 40 responded, but by August, 1920, 18 of those companies had resumed production, with only 22 still idle. *Williamson Daily News,* August14, 1920; "Testimony of C. F. (Frank) Keeney," *West Virginia Coal Fields,* 124; Source for Letter from Keeney to Governor Morgan, July 11, 1921, read into the record by Keeney: "Testimony of C. F. (Frank) Keeney," *West Virginia Coal Fields,* 107; Ephraim F. Morgan to S. D. Stokes, August 11, 1921 and Stokes to Morgan, August 12, 1921, Stokes Papers, WVRHC; Source for Exchange between S. B. Avis and Keeney: "Testimony of C. F. (Frank) Keeney," *West Virginia Coal Fields,* 124.

48 Mooney, *Struggle in the Coal Fields,* 53, 69–70; N. A. MacKenzie to John J. Cornwell, December 3, 1920, Cornwell Papers, WVRHC; George C. Wolfe to Justus Collins, September 21, 1921, Collins Papers, WVRHC.

49 See author's dissertation chapter nine for more on the interaction between the union, Sheriff Blankenship, Mayor Testerman, and Blankenship's successor A. C. Pinson; Source for Republican reaction: *Mingo Republican,* June 3, 1920; District 17 vice-president William Petry and Mother Jones were both NPL supporters, in fact Petry was the head of the West Virginia NPL; Source for Petry as UMWA officer: Mooney, 56; Source for Petry in NPL: Harris and Krebs, 176; Source for Mother Jones: Winding Gulf Coal Operators' Association "Bulletin U," 30 August 1920, Collins Papers, WVRHC. The NPL advocated for increased rights for industrial workers, an end of monopoly, nationalization of the transportation industry, and a steeply graduated income tax." Stock, 75; Source for African Americans and Montgomery: *Mingo Republican,* May 15, 1920; Source for Mingo endorsement: *Williamson Daily News,* June 4, 1920; Source for African-

American loyalty: "Negro Activity," October 23–30, 1920 in *Surveillance*; For more on the anti-union activities of McDowell's black leadership see: Trotter, "Black Miners," in *The United Mine Workers of America: A Model of Industrial Solidarity?* edited by John H. M. Laslett, 291–292; Source for Damron-Brewer relationship: "Testimony of Sid Hatfield," *West Virginia Coal Fields*, 216; "Testifies Hatfield Killed Mayor," *New York Times*, May 25, 1921; "Testimony of C. E. Lively," *West Virginia Coal Fields*, 390–391.

50   Hatfield correspondence, letter no. 29; Conditions described by Matewan Oral History Project narrators included: lying on the floor of passenger train cars to avoid gunfire (Vicie Blackburn interview) and hiding in basements for weeks, also to avoid stray bullets (Bertha Staten interview); Circuit court judge R. D. Bailey observed to a reporter, "if the violence were to cease, public sentiment would compel the coal operators to accept unionism." Arthur Warner, "Fighting Unionism with Martial Law," *Nation* 113 (12 October 1921): 396; "Testimony of H. C. Ogden," *West Virginia Coal Fields*: 944–948, 946. Ogden, the publisher of the *Wheeling Intelligencer*, was one of the leading proponents of the industrial democracy movement in West Virginia; Batteau, 125. For more on the press and national public opinion see Batteau's Chapter 6 "Which Side Are You On?," and also Jeffreys-Jones, 34; By the time of the Massacre trial, the union had spent over one hundred thousand dollars on strike-related legal expenses. Between 1920 and 1922 the UMWA's total legal expenses exceeded eight hundred thousand dollars. Dubofsky and Van Tine, 79. Source for Miners' attacks on: replacement workers: *Williamson Daily News*, 14 December 1920; Source for law officers, e.g. prohibition officer Harry Staten: "State of West Virginia v. J. S. (Calvin) McCoy," *Reports of the West Virginia Supreme Court of Appeals* 91 (April 25, 1922–October 17, 1922): 262–268; Source for private citizens, e.g. Ance Hatfield: *Bluefield Daily Telegraph*, 10 September 1920; Mingo's local union attorney, Thomas West was disgusted particularly by the Massacre defendants' efforts "to frame up their own defense." Report of #9, 25 June 1920, Lewis Collection, ERCA. West, who like District 17's lead attorney Houston, was also a Socialist, had worked as a miner until age 21. "Thomas West," in *Progressive West Virginians* (1923): 235. West also represented the striking miners who attacked the Hungarian co-operative mine and miner. See author's dissertation chapter nine. Union organizer J. L. Workman confided to another secret agent that Frank Keeney gave the striking miners at Matewan whatever they wanted because they knew details about the "Matewan

murders" that could cause "the organization a great deal of harm." Report of #5, 19 February 1921, Lewis Collection, ERCA; "Testimony of A. E. Hester," *West Virginia Coal Fields*: 802–838. Hester, who had helped buy rifles for miners with unused relief funds grew increasingly disenchanted with what he considered the misuse of the weapons. When he announced his intention of leaving the strike zone and returning to Charleston, he discovered that District 17 had accused him of embezzling strike funds. After surrendering to the Kanahwa County sheriff, he spent 41 days in jail, despite the absence of an indictment; Source for Lewis in West Virginia: Philip S. Foner, *Women and the American Labor Movement: From Colonial Times to the Eve of World War I* (New York: The Free Press, 1980), 244; Source for Lewis at treason trial: Lunt, 159; C. F. (Frank) Keeney to Secretary of War Newton Baker, October 14, 1920, copy sent by #19 to Governor Cornwell, Cornwell Papers, WVRHC. In the letter Keeney quotes an observation from *Coal Mining Review* 15 September 1920, concerning the correlation between the arrival of federal troops and the demise of strikes. See also Fagge, 139.

51   Corbin, *Life, Work, and Rebellion*, 99; Lunt, 151; *Mingo Republican*, June 3, 1920; "Testimony of Jett Lauck," *West Virginia Coal Fields*: 1036–1037 and "Testimony of Samuel Untermyer," in same, 697–719; S. D. Stokes to James P. Woods, July 2, 1921, Stokes Collection, WVRHC.

52   Source for Union's strike expenditure: Hennen, *Americanization*, 112 and Lewis, *Black Coal Miners*, 163; Source for end of strike: *New York Times*, October 28, 1922; Mooney, *Struggle in the Coal Fields*, 124–127; Source for Lewis actions: Melvyn Dubofsky and Warren Van Tine, *John L. Lewis: A Biography* (New York: The New York Times Book Company, 1977), 81; Source for what happened to Keeney and Mooney: Mooney, *Struggle in the Coal Fields*, 127–128; Melvin Triolo, interview with John Hennen, Summer 1989 Matewan Oral History Project; Source for what happened to Blizzard: Cabell Phillips, "The West Virginia Mine War," *American Heritage* 25 (August 1974): 58–61, 90–96, 94; Mooney, *Struggle in the Coal Fields*, x; "William (Bill) Blizzard," *West Virginia Heritage Encyclopedia*, vol. 3: 470–471.

53   Conley, *History of the West Virginia Coal Industry*, 262; David E. Whisnant, *Modernizing the Mountaineer: People, Power, and Planning in Appalachia* (Knoxville: University of Tennessee Press, 1994), 110.

54   Justus Collins to George C. Wolfe, September 16, 1920, Collins Papers, WVRHC; *Bluefield Daily Telegraph*, July 2, 1920; F. A. Lindsay, on behalf of

Allburn Coal & Coke Company, to S. D. Stokes, December 21, 1920, Stokes Papers, WVRHC; S. D. Stokes to James P. Woods, July 2, 1921, Stokes Papers, WVRHC; M. Z. White to Greenway W. Hatfield, August 28, 1921, Ephraim F. Morgan Papers, WVRHC; S. D. Stokes to G. W. Coffey, August 19, 1920, Stokes Collection, WVRHC; Source for operators and officials indicted for bribery: J. M. Tully, W. A. Wilson, and G. R. C. Wiles, *Mingo Republican*, July 8, 1920; Source for agent provocateur: C. F. Keeney to Newton Baker, October 4, 1920, Cornwell Papers, WVRHC; "Statement of R. H. Kirkpatrick," *West Virginia Coal Fields*, 101; "Testimony of J. R. Brockus," *West Virginia Coal Fields*, 344–345.

55 Source for timing of raid: "Testimony of Frank Ingham," *West Virginia Coal Fields*, 34; Source for martial law case: "Ex Parte Lavinder, et al," *Reports of the West Virginia State Supreme Court of Appeals*, 88 (February 22, 1921–June 24, 1921): 713–721, 721; Source for Breedlove's depiction by operators: "Testimony of Albert E. McComas," *West Virginia Coal Fields*, 303; Source for union: "Keeney Exhibits, #18 and 19," "Testimony of C. F. (Frank) Keeney," *West Virginia Coal Fields*, 166–167; *Mingo Republican*, May 23, 1913. "Keeney Exhibit #18"; Major Thomas B. Davis to Ephraim F. Morgan, June 5, 1922, Morgan Papers, WVRHC; Heber Blankenhorn, "Marching Through West Virginia," *Nation* 113 (September 14, 1921): 288; Source for definition of Flag Day: June 14 "is the anniversary of the adoption (1777) of the national flag of the United States," *Webster's Encyclopedic Dictionary*, 356; "Testimony of Albert E. McComas," *West Virginia Coal Fields*, 304–305; *New York Times*, May 27, 1921; Winding Gulf Coal Operators' Association, Bulletin #62, May 31,1920, Collins Papers, WVRHC; Source for date of meeting: *Coal Trade Journal* 51 (June 2, 1920): 588.

56 Justus Collins to George C. Wolfe, April 16, 1920 and "Bulletin F," May 21, 1920, Justus Collins Papers, WVRHC; "Testimony of R.C. Kirk," *West Virginia Coal Fields*, 471; "Testimony of W. R. Thurmond," *West Virginia Coal Fields*, 867; "Testimony of C. [sic] T. Blankenship," *West Virginia Coal Fields*, 487; McDowell County sheriff S. A. Daniel to John J. Cornwell, September 2, 1920, Cornwell Papers, WVRHC; William N. Cummins to John J. Cornwell, November 17, 1920 and G. T. Blankenship to same, November 6, 1920, Cornwell Papers, WVRHC; Lee, *Bloodletting*, 67; *Charleston Gazette*, August 2, 1921, quoted in Corbin, *Life, Work and Rebellion*, 210.

57 Neil Burkinshaw, "Labor's Valley Forge," *Nation* 110 (December 8, 1920): 639; Report of #5, February 19, 1921, Lewis Collection, ERCA; *Bluefield Daily*

Telegraph, July 2, 1920; *Coal Trade Journal* 51 (June 2, 1920): 588; Source for notices of dissolution: *Williamson Daily News*; Sources for publication, S. D. Stokes to Thomas B. Garner, May 26, 1920; "Testimony of E. L. Bailey," *West Virginia Coal Fields*, 277, 284; "Strike Activity" 9 October 1920, *Surveillance*; The quotations are borrowed from: "West Virginia's War," *Literary Digest* 67 (December 18, 1920), 16.

58   William K. Klingaman, *1919: The Year Our World Began* (New York: Harper & Row, 1989 [1987]), 548–549; and Robert H. Wiebe, *The Search for Order: 1877–1920* (New York: Hill and Wang, 1967), 293–302, 288–289; Hennen, *Americanization*, 1–2, 69. See Chapters 1–2 of *Americanization* for a synopsis of both the national and West Virginia application of the concept of engineered consent; See Murphy, *World War I and the Origin of Civil Liberties*, 57, 82, 101–102, and 271; Bernstein, 75; Hinrichs, 181.

59   Source for Cornwell: "Resign As Governor Whenever Unable To Preserve Order," *Bluefield Daily Telegraph*, February 7, 1920; Source for Morgan: "Governor Explains Why Federal Troops Brought Into West Virginia—Six Thousand Deluded Men In Insurrection, Reason He Offers (New York Commercial, 1921 [reprint]), Pamphlet 7640, WVRHC; Chapter 69 "The Jury Bill," Senate Bill #14, *Acts of the West Virginia Legislature for the Year 1921*, 183–184; S. D. Stokes to S. B. Avis, March 22, 1921, Stokes Papers, WVRHC; "State of West Virginia v. J. S. McCoy," *Reports of the West Virginia State Supreme Court of Appeals* 91 (April 25, 1922–October 17, 1922): 262–268; *Bluefield Daily Telegraph*, October 1, 1921; J. C. McWhorter, "Abolish the Juries," *West Virginia Law Quarterly* 29 (January 1923): 97–108; Senate Bill# 359, *Acts of the West Virginia Legislature for the Year 1921*, 322–329; *United Mine Workers Journal* 33 (October 1, 1922): 3; "Testimony of William McKell," *West Virginia Coal Fields*, 941; Source for Pond Creek: Lane, *Civil War*, 22; Source for Borderland and Red Jacket: Lunt, 96, 152–153, 172, 166–167, 145–149, and 154.

60   James M. Cain, "The Battleground of Coal," in *The West Virginia Mine Wars: An Anthology*, edited by David A. Corbin (Charleston, WV: Appalachian Editions, 1991): 151–159, 155. The article originally appeared in the *Atlantic Monthly*, in October 1922; Source for President Harding: *The United Mine Workers in West Virginia*, 6; *Personal Views of Senator Kenyon*, 6; Clayton Laurie, "The United States Army, The Return to Normalcy in Labor Dispute Interventions: The Case of the West Virginia Coal Mine Wars, 1920–1921," *West Virginia History* 50 (1991): 1–24,

8, 15–18; "Strikes," October 23, 1920, *Surveillance*; Edward Eyre Hunt, F. G. Tryon, and Joseph H. Willits., eds., *What The Coal Commission Found: An Authoritative Summary By The Staff* (Baltimore: The Williams & Wilkins Company, 1925), 34.

61   Bernstein, 190–191. For more on labor law in this period see: Bernstein, *The Lean Years*, Chapter 4 "Labor v. The Law"; Arthur Gleason, "Company-Owned Americans," *Nation* 110 (June 12, 1920): 794–795, 795; Source for "puttering": *Bluefield Daily Telegraph*, January 28, 1922; *Personal Views of Senator Kenyon*, 4; Johnson, *Politics of Soft Coal*, 111; Source for "desolate": Arthur Gleason, *The Book of Arthur Gleason: My People, and "A.G." An Appreciation By Helen Hayes Gleason* (New York: William Morrow & Company, 1929), 183.

62   Margaret Hatfield interview. This comment was made off-tape, but the interview does contain Ms. Hatfield's memories of Clare Overstreet, the subject of her statement.

63   As in other stories of violence, the Massacre's impact on subsequent events resulted from its utility as a source of propaganda. As historian Jane Dailey observed of the 1883 Danville, Virginia race riot, the violence of that day influenced the 1883 Virginia election because of how Democrats and Readjusters portrayed the causes of the event. See Dailey, "Deference and Violence," 581–582; David Fowler and David Robb, "Statement of Conditions in Mingo County, W.Va.," *United Mine Workers Journal* 31 (January 1, 1921): 10; *National Coal Mining News*, editorial, September 29, 1921. Wightman Roberts, the editor of the *Mining News* corresponded frequently with Governor John J. Cornwell; Jeffreys-Jones, 37 and 155–157; Charles Frederick Carter, "Murder to Maintain Coal Monopoly" *Current History* 15 (1922): 597–603; The article's title refers to what Carter asserted was the UMWA's "twenty-three years of arson, assault, and assassination in West Virginia"; Batteau, 124, 115. Batteau asserts that the near constant invocation of the Feud resulted in the construction of "feud-feudal homonym." An example of the homonym can be found in a 1920 issue of the *Literary Digest*. "West Virginia's War," *Literary Digest* 67 (December 18, 1920): 16–17. The subject bullet at the top of these pages reads "Feudalism in America," while the article refers to the strike as West Virginia's bloodiest feud.

64   For more on Harlan County, Kentucky miners see: George J. Titler, *Hell in Harlan* (Beckley, WV: BJW Printers, n.d.); *What the Coal Commission Found*, 235–236; *New York Times*, May 27, 1921, 3. The *Times* actually used the phrase "concentration camp" but in the post-Holocaust world the phrase has an em-

phasis that must be invoked carefully; in the author's opinion "camp" was suggestive enough; Source for The LaFollette Commission: Batteau, 123; Source for Blankenhorn as chief investigator: Beik, 267; Spivak, 59; Lunt, 181.

65   Source for "demons": Richard Burgett interview, *On Dark and Bloody Ground*, 108; Hatfield correspondence, letter no. 29; Sources for impact of events, "ugliness": Hiram Phillips interview with John Hennen, Summer 1989 Matewan Oral History Project; Virginia Grimmett interview; Venchie Morrell (1990) interview; Hawthorne Burgraff interview; See Velke, 200 and Paul Lively interview; Johnny Fullen interview; Another positive outgrowth of the rediscovery of the Matewan Massacre has been the willingness of descendants of those who have been vilified to finally come forward and tell their story. The Matewan Oral History collection now includes an interview with the son of C. E. Lively. Also, Dick Redden, the grandson of Albert Felts, has visited the Eastern Regional Coal Archives. In a local newspaper interview, Mr. Redden explained that his mother (Felts's daughter) never discussed his death, and as a result Mr. Redden never learned of his family's "role in the coalfields until 1982." *Bluefield Observer*, 13 May 1992.

6

CONCLUSION: THE MATEWAN MYTH

IN 1978, Henry D. Shapiro's *Appalachia On Our Mind* challenged scholars to consider "what problem [the study of Appalachian history] solves and whose interests—intellectual as well as practical—it thereby serves." Drawing inspiration from this parting thrust of Shapiro's ground-breaking monograph, *Matewan Before the Massacre* has sought to explore the "back story" of the events of May 19, 1920. In the process, it was discovered that for decades journalists and historians have constructed and cultivated a version of the Matewan Massacre that focused solely on the struggle between "the union" and "the company." This simple dialectic left no room for an examination of the human agency of the people of Matewan and Mingo County. From the reporters who witnessed the events set in motion by the Massacre, to the historians seeking to understand Appalachia's rural-industrial working class, those who wrote of the Massacre edited and defined it to fit into the larger picture of American working class activism. Why the Massacre occurred in Matewan and what were its local origins were questions left out of analyses of the broader struggle for workers' rights in the region's anti-union coalfields.[1]

Within days of the Massacre, journalists arrived in Matewan titillated by "vague ideas of romance, adventure and bravour," and by the inherent danger of investigating industrial atrocities in a "lawless section" already made infamous by the Hatfield-McCoy feud. Since most had spent their professional lives in the urban centers of the Eastern seaboard, the reporters who came to West Virginia between 1920 and 1922 possessed little first-hand knowledge of Appalachia. Therefore, their impressions of Matewan and Mingo County were shaped by the 50 years of local-color literature that had portrayed the upcountry South as a place in, but not of, America. They described the men of Matewan as being "of an inheritance and habit apart," and "members of an atrophied race, a weaker strain of American stock." Also overwhelmed by the events they witnessed, the reporters lost their objectivity and were swept up into what they perceived as a fight against the injustices perpetrated by industrial capitalism. "Oblivious to the historical and political causes" that undergirded the union-company struggle, the reporters transformed the miners of Mingo County and southern West Virginia into faceless "people who had lost a culture" and were now imprisoned in an unending string of rural ghettos.[2]

When it became evident that the miners were losing the battle to unionize Mingo, they were dismissed as "flotsam cast up by the backwash of a mighty struggle, pathetically loyal to a cause" they never really understood. Disheartened, the journalists retreated to New York and moved on to other causes. Arthur Gleason became the vice-president of the League for Industrial Democracy. Heber Blankenhorn worked with John Brophy on the UMWA's Nationalization Research Committee. Jack Spivak accepted a commission from the *Call* to cover the Sacco and Vanzetti trial. Although the memory of Matewan and the West Virginia mine wars remained with them, only one seemed to grasp the impact of the press coverage on the lives of those whom the reporters sought to help. Shortly before his death in 1923, Arthur Gleason lamented that "the brilliant crusading of [these] outsiders" had been "attended by unwitting damage to obscure people." The journalists who wrote about Matewan, Mingo County, and the Williamson-Thacker strike immortalized the UMWA's struggle to unionize southern West Virginia, but in so doing, they diminished both the humanity and the Americanness of the people at the epicenter of the contest.[3]

However, the reporters should not bear alone the blame for either the "othering" of the people of Matewan and Mingo County and southern West Virginia or the loss of a localized focus to the story. In the absence of a local consensus regarding the Massacre and strike, much less their root causes, explanations were gladly provided by others interested in the outcome. As a result, the United Mine Workers of America moved to the foreground of the narrative, while Mingo County and its inhabitants were pushed to the background.[4]

Representatives of the UMWA also set the tone for the denigration of Mingo's miners. Although black miners spoke most eloquently about the constitutionality of their struggle, during the senate hearings, the union's spokesman stressed the whiteness of the county's mining population in order to underscore the "Americanness" of their struggle. When challenged to explain strike-related violence in West Virginia, both District 17 President Frank Keeney and the union's lead attorney, Harold W. Houston, invoked the feuding mountaineer stereotype. In seeking to underscore the union's ability to prevent such strife, District 17 Secretary Fred Mooney purposefully likened the Williamson-Thacker strike to the 1912–1913 Paint Creek and Cabin Creek strike. Because Mooney's observation was offered before a U.S. Senate committee chaired by the same senator who had led the 1913 investigation, his intent was clear—to convince the federal government that unionization was the only cure for the periodic disruption of the state's coal production.[5]

The intervention of federal power on behalf of "law and order" allowed the operators to win the 1920–1922 mine war. Intoxicated by the success of driving the union out of Mingo County, the operators promptly launched a propaganda campaign in which they claimed that because of West Virginia's industrialization, little poverty existed in the state. West Virginia governor Ephraim Morgan excoriated the UMWA for fomenting rebellion at the behest of radical agitators. The editor of the industry magazine, *Coal Age*, declared that "the name of Mingo has become [more than] . . . a household word . . . the name of Mingo has become a stench in the nostrils of good American citizens and the source of that malodorous smell bears the union label." Even contemporary academics' criticism of the union's actions in West Virginia appeared to endorse the operators' claim that the operators had been the defenders of the public's interest. Not until New Deal legislation formally endorsed the UMWA's legitimacy did southern West Virginia's operators realize that although they had won the 1920–1922 battle, they had lost the war with the union.[6]

The union's ascension in the 1930s was accompanied by a more subtle victory. From that time until recently, the UMWA has been portrayed as the proactive, and eventually ennobled, agent in the organization of West Virginia's and Appalachia's working class. Events such as the Matewan Massacre were mentioned but depicted as only one of a series of "'outbreaks of violence' that were exclamation points to the economic hardships of the times and the operators' unrelenting campaign to cut wages and break the union." Following the rediscovery of Appalachia during the turbulent 1960s, historians inspired by the New Left critique of American capitalism turned their attention to the equally volatile history of West Virginia's coalfields. In their rush to celebrate the miners' challenge to the status quo, recent scholars have also naively assessed the past through the lens of their own era and forgotten this lament from the autobiography of Malcolm Ross, author of *Machine Age in the Hills*: "I tried to show how the impersonal thread of economic interest can tangle tragically with the threads of human lives . . . without anyone being egregiously at fault."[7] By analyzing the West Virginia mine wars and the Matewan Massacre in particular solely as events in labor history, historians have forged a new cognitive dissonance, whether knowingly or unwittingly. Like the contemporary observers criticized by Allen Batteau, historians of the mine wars ignored the other historical and political causes that influenced the miners' struggle, and thus unintentionally perpetuated the cycle of elevation through denigration first begun by the UMWA and the journalists of the 1920s. By focusing on the leadership of District 17, David Corbin, the pioneer of the new interpretation, fell into the trap of the "Great Man" methodology and turned the miners into "essentially passive" victims in need of liberation. In an otherwise excellent explication of the legal ramifications of the West Virginia unionization struggle, Richard Lunt contextualized the miners' violence simply by labeling Matewan and Mingo County as "Hatfield-McCoy country." As if following a cue left behind by the contemporary observers of the events of 1920–1922, historians have collectively focused their "discursive efforts" on southern West Virginia as a single entity with little regard for the unique experiences and motivations of "the historical actors" of Mingo County.[8]

The resulting paradigm of the non-union coalfields of southern West Virginia created "not only knowledge but also the very reality [it appeared] to describe." Don Chafin's dictatorial rule of Logan County symbolized the operators' subversion of the political autonomy of southern West Virginia's coal counties but also precluded a discussion of the impact of Mingo County's unique political volatility, which, for example, allowed coal miner Elihu Boggs to serve as mayor of Matewan and clerk of the county court or placed the pro-union G. T. Blankenship in the sheriff's office in 1920. The isolation of the coal camps in the more developed coalfields and the preoccupation of historians with the company system have prevented an examination of the economic and political tensions fostered by the coexistence of the "company system" and independent merchants in communities such as Matewan. The emergence of an interracial working-class solidarity was ascribed to the racial and ethnic heterogeneity of the mining camps, while the actual low number of African Americans and European immigrants in Mingo and the white miners' bigotry was ignored. The scholars' pro-union bias and macrocosmic representation of southern West Virginia has ultimately led to the inclusion of factual errors, as scholars have increasingly relied on their predecessors' conceptualizations when examining the available body of evidence. Just as Herbert Gutman noted that labor historians had erred in "[spinning] a cocoon around American workers, isolating them from their own particular subcultures and the larger national culture," scholars of the mine wars have diminished the dignity of the people by reducing the coal struggle to the noble union fighting the evil companies to liberate victim-miners.[9]

Because historians persisted in confining the story of the Matewan Massacre within the boundaries of Appalachian labor study, the history of the Massacre still exemplifies an assertion made by Christian missionary John C. Campbell in 1921: "This is a land . . . about which perhaps more things are known that are not true than any part of the country." By restricting the parameters of the conflict to the struggle between the miners and the operators, historians also have constructed a distorted historical narrative that allows other Americans to continue trivializing the experiences of the people of Matewan and Mingo County. A contemporary spectator of the Williamson-Thacker strike presciently predicted the cumulative effect of the historiography of the second mine war. As James M. Cain noted, fellow Americans flock to Matewan hoping to find the notorious Devil Anse or Sid Hatfield but instead find "the Last of the Mohicans turned tourists' cook." Offered the enticement of capitalizing on outsiders' interest in their "history," the people of Matewan and Mingo County have been expected to remain frozen in time, caretakers of an historical place maintained to remind the rest of America of "how far the nation has come from its essential self."[10]

Verification that the people of Matewan know that they have been "oth-ered" can be found in the Matewan Oral History project interviews from the summers of 1989–1990. Although the interviews offer tantalizing glimpses of the Feud and the Massacre, they also demonstrate the com-munity's ability to synthesize apparently contradictory actions and events to a degree that has eluded a score of academicians. When Venchie Morrell spoke kindly of both Sid Hatfield and P. J. Smith, the superintendent of the Stone Mountain mine, despite witnessing an incident during which Hatfield assaulted Smith, he invited his interviewer to reexamine the his-torical record and character of this coal company official in Mingo County. Instead of discovering evidence of heavy-handed mistreatment of his em-ployees, she found evidence that not only was Smith respected by the men who worked for him; he was also a casualty of the industrial struggle.[11]

By restoring the human complexity of those caught up in the Massacre and strike, the oral history project interviews fleshed out people who had been reduced to archetypal caricatures. Who could imagine that a man who stalked the streets of Matewan with murderous intent in the Massacre's aftermath would later tearfully beg his firstborn, a daughter, "not to ever move away and leave him"? Or that this same individual would befriend a man he would have gladly killed on May 19, 1920, partner with him in the mines, and upon their deaths lie in the same cemetery? Unlike the incident related by Venchie Morrell, the veracity of this last story could not be established. Given the subsequent religiosity of the man who first told it, the tale of friendship between two former enemies might have been an instructional parable concocted, however sincerely, for his chil-dren. However, cynics tempted to dismiss the story as a myth should be cautioned by this observation by historian Sean Wilentz: "Myths after all, draw on popular beliefs and . . . once formulated, they help order people's understanding of the world and tell us something about social relations." The story of the miner and the repentant Baldwin-Felts agent symbolizes the resilience of the people of Matewan, who perceived the need to make peace with their past and move forward.[12]

As a document of communal expression, the 1989–1990 interviews also contain an unstated poignant plea that Matewan's history be seen as more than just a string of violent incidents and notorious characters. Once motivated to share their memories, the Matewan narrators exhibited little interest in proving that they knew who shot Mayor Testerman or who fired the Massacre's first shot. Rather, they wished to show that outsiders' preoccupation with the event had created a distorted historical narrative that culturally isolated them and diminished their status as Americans. What they could not know was that they had inspired a historian to explore ways of affirming their desire to integrate the Massacre and strike into the broader continuum of the history of their community. That was the genesis of this study.

Matewan Before the Massacre demonstrates that the story of the Massacre, unshackled from the rhetoric of outside interests and observers, is, at its core, an American story, which captures a pivotal moment in our national history. In the late nineteenth and early twentieth century, political machines and reformers vied for power in communities across the nation, not just in Appalachia. Industrialization reconfigured the economic and social landscape of the entire country, not just southern West Virginia. Domination through violence occurred not only on the streets of Matewan, but also in the tenement districts of New York, the cotton fields of the South, and the docks of the Pacific Northwest. Still, rarely has a community been so enslaved to an externally imposed version of its past as the town of Matewan.[13]

If liberated from the shadow of the Massacre myth, the oral histories of Matewan could reveal hundreds of stories which would resonate with Americans from all regions and all walks of life. Joe Morrell, the only man blackballed for his union activism, arrived in America from Italy with five brothers whom he never saw again after departing New York for West Virginia. He married Sophie, an Austrian immigrant who delivered twins in the winter's cold, sheltered only by the thin canvas of a union tent. They reared a son, who along with other children, mocked the local Klansmen who marched through Matewan, their identities betrayed by their shoes and the way they walked. The son of Joe and Sophie grew to manhood and joined the other sons of Matewan and went off to fight for his country in the second World War.[14]

How many American families can recount the initial sighting of the Statue of Liberty by an ancestor? Would they not empathize with Manuel Barrios, the son of a Spanish mason recruited by Red Jacket in the midst of the Williamson-Thacker strike, who 60 years later could still vividly recall his first glimpse of "the beautiful lady?" Should his story of learning English by shopping for his mother at the company store be forgotten because his father was a "scab"?[15]

If the stories of the Morrell and Barrios families reveal another side of Matewan, the story of the Collins' family encapsulates the tragic injustice of isolating Appalachian history. John and Steve Collins, two of the few men sent to prison for Mingo's strike-related violence, held onto their faith in their union and in their country. Steve sent a son off to war and lost him the day before Christmas Eve, 1944. John's grandson survived the killing fields of Vietnam, only to die at age 21 in a mining incident two months after his return from war.[16]

Many other stories from Matewan mirror the poignancy of those described above. The youth of Matewan need to hear them, not just the stories of the Feud and the Massacre. They need to hear them because, like their forebearers, they also are caught up in the ongoing struggle over West Virginia's economic future, and they too will be judged for the decisions they make. In the classrooms of Mingo County today, the sons and daughters of union families again sit beside the children of non-union families. Although economic alternatives to mountaintop removal remain underexplored, one form of economic viability lies in heritage tourism. Still, must the youth of Matewan and Appalachia sacrifice their history to myth and stereotype to earn their bread? If there is any hope that the political, economic, and cultural cycles that have kept their communities divided and impoverished can be broken, a reconceptualized and demythologized history must be written. It is my hope that perhaps, however infinitesimally, *Matewan Before the Massacre* will assist one community in this enterprise and help remind historians that all history is local.[17]

NOTES

1   Henry D. Shapiro, *Appalachia On Our Mind: The Southern Mountains and Mountaineers in the American Consciousness, 1870–1920* (Chapel Hill: University

of North Carolina Press, 1978), 265; E. P. Thompson, *The Making of the English Working Class* (New York: Vintage Books, 1966), 9; Gutman, 3–78, 53; "Edited" and "defined" were borrowed from Edward Said, whose intellectual history of Western Europe's response to Eastern culture provides a framework for the study of the phenomenon of "othering," the means by which one culture is defined through contrast with another. Edward W. Said, *Orientalism* (New York: Pantheon Books, 1978), 167, 54.

2   Source for "vague ideas": "Baby Footprints in the Snows of Mingo County," *United Mine Workers Journal* 32 (January 15, 1921): 14–15, 14; Source for "lawless section": "A. G." by Helen Hayes Gleason, in *The Book of Arthur Gleason*, 99; Source for influence of local color writers: Shapiro, 18; Source for "inheritance apart": "The Primitive Mountaineer" *New York Times*, August 3, 1921; Source for "weaker strain": Cain, "Battleground of Coal," *Mine Wars Anthology*, 154; Source for lost objectivity: Spivak, 110 and Mildred Morris, *West Virginia Federationist*, September 15, 1921; Source for "oblivious": Batteau, 125; Sources for lost culture: Ross, *Death of a Yale Man*, 94–95 and Lane, *Civil War in West Virginia*, 21–22. Lane's simplified depiction of Mingo's communities in 1920–1922 mirrors the representation of the Tug Valley by journalist T. C. Crawford, who wrote the first popular account of the Hatfield-McCoy feud in the late 1880s. Both Lane and Crawford painted a picture of the local community that excluded any elements that would have been recognizable to their contemporaries in other parts of the country. For a synopsis and analysis of Crawford's *An American Vendetta*, see Waller, *Feud*, 7–8.

3   Source for "flotsam": Cain, "Battleground of Coal," in *Mine Wars Anthology*, 154; Ross, 106; Gleason, *The Book of Arthur Gleason*, 202, 204; Coleman, 106; Spivak, 131.

4   The ability of people from the same community, possessing essentially the same values, to "narrativize" an event differently is universal. See Ann Kane, "Reconstructing Culture in Historical Explanation: Narratives as Cultural Structures and Practice," *History and Theory* 39 (October 2000): 311–330, 317–318. Kane studied the Irish Land War of 1879–1882. The co-option of an event's meaning and importance is also examined in Dailey, "Deference and Violence," 581–582.

5   Source for black miner: "Testimony of George Echols," *West Virginia Coal Fields*, 470; Source for spokesman: "Opening Statement of Neil Burkinshaw," *West Virginia Coal Fields*, 7; "Testimony of C. F. "Frank" Keeney, *West Virginia*

Coal Fields, 184; Houston Brief, 33; "Testimony of Fred Mooney," *West Virginia Coal Fields*, 19.

6 Phillips, "West Virginia Mine War," 94. Phillips was the son-in-law of Frank Keeney; "'The Truth About West Virginia,' an address by Randolph Bias, an attorney of Williamson, W. Va., to the Kiwanis Club of Portland, Oregon on September 12, 1922," (Williamson, WV: n.p., 1923), 13. P4344, Pamphlet Collection, WVRHC; "Governor Explains," P7640, Pamphlet Collection, WVRHC; "Is Insurrection Proof of Americanism?" *Coal Age* 52 (September 7, 1921), editorial page; See Hinrichs, *The United Mine Workers of America and the Non-Union Coalfields*, (1923), and Mary Hurst, "A Social History of Logan County, West Virginia, 1765–1928" (Master's thesis, Columbia University, 1928); Phillips, 94; Mooney, 164.

7 Corbin, Introduction, *Mine Wars Anthology*, ii; "John L. Lewis and the International Union of the UMWA: The Story from 1917–1952," (International Executive Board of the UMWA: Washington, D.C.: 1952), in the John L. Lewis Papers, State Historical Society of Wisconsin; Ross, 106.

8 In *Appalachia On Our Mind*, Shapiro defines cognitive dissonance as the contrast between perception of reality and the reality that actually exists. Shapiro, xvi; Batteau, 125; Critique of Corbin's "Frank Keeney Is Our Leader," by Warren Van Tine and John W. Hevener in *Essays in Southern Labor History: Selected Papers from the Southern Labor History Conference*, 157; Lunt, 146; Quoted phrases borrowed from Billings and Blee, 16.

9 Said, 94; Source for contemporary depiction of monolith: Arthur Gleason, "Company-Owned Americans," *Nation* 110 (June 12 1920): 794–795. In explaining how operators dominated southern West Virginia's coal counties, Gleason discusses only Logan, and its sheriff, Don Chafin; Source for historian who continued the depiction: Corbin, *Life, Work, and Rebellion*, 115–116, 210–235. Like Gleason, Corbin names only Chafin and discusses at length his corrupt and oppressive regime, but does not mention Blankenship; Source for contemporary depiction of coalfields: Lane, *Civil War in West Virginia*, 21–22; Sources for historians who continued depiction: Corbin, *Life, Work, and Rebellion*, 68 and Fagge, 210–211; 1912–1913 *West Virginia State Gazetteer and Business Directory*, 469; and author's dissertation chapter 10, note 50; Source for ethnic misrepresentation by historian: Skeen, abstract; Source for contemporary representations: *New York Times*, December 7, 1920; Joe William Trotter's essay, "Black Miners in West

Virginia," explains that Mingo's black miners suffered a "disproportionate" share of the operator-directed reprisals. The essay can be found in *The United Mine Workers of America: A Model of Industrial Solidarity?*, 289; The only documented case of permanent exclusion ("blackballing") from mine employment in Mingo County involved Joe Morrell, an Italian-born miner who spoke several European languages. Because of his linguistic facility, Morrell would meet the "transportation" trains of replacement workers and in their native tongues reveal that they were entering a strike district. Venchie Morrell 1989 and 1990 interviews; The most egregious example of factual error appears in Roger Fagge's comparative study of political culture in southern West Virginia and South Wales. Fagge notes that because the operators were so powerful, the Baldwin-Felts agents involved in the Massacre and the deaths of Sid Hatfield and Ed Chambers never faced legal judgment. Fagge, 146. As this author has documented, although the operators' power allowed them to subvert the law to protect the agents, they assiduously observed the form of the law by not interfering with the trial of either the agents who survived the Massacre or the ones who assassinated Hatfield and Chambers. See author's dissertation chapter ten. Although more eminent scholars may disregard Fagge's study because of its outdated premise—an exploration of the failure of West Virginia miners to develop a political radical class consciousness—his novel approach to analyzing the mine wars will ensure its inclusion in the historiography; Gutman, *Work, Culture and Society*, 11.

10   John C. Campbell, *The Southern Highlander and His Homeland* (Lexington, KY: University Press of Kentucky, 1969 [reprint of 1921 edition published by the Russell Sage Foundation]), original foreword, xxi. *Southern Highlander* was published two years after Campbell's death in 1919; "Trivializing" is borrowed from Altina Waller, "Feuding in Appalachia: The Evolution of a Cultural Stereotype," in *Appalachia in the Making: The Mountain South in the 19th Century*, edited by Dwight Billings, Mary Beth Pudup, and Altina Waller (Chapel Hill: University of North Carolina Press, 1995), 370; James M. Cain, "West Virginia: A Mine-Field Melodrama," in Corbin, *Mine Wars Anthology*: 144–150. This essay first appeared in *Nation* 113 (July 27, 1923); Shapiro, 261: Durwood Dunn, *Cades Cove: The Life and Death of a Southern Appalachian Community, 1818–1937* (Knoxville: University of Tennessee Press, 1988).

11   Author's assertion based on an ethnographic analysis of Matewan Oral History project interviews for the seminar paper, "Lost Voices: Memories of the

Matewan Massacre," written in the spring of 1994; Venchie Morrell, 1990 interview; Sources for contemporary documentation of Smith's behavior: James P. Woods to S. D. Stokes, May 21, 1921, Stokes Collection, WVRHC; and "Matewan Firebugs Burn Mine House, Firemen Refuse Aid," *New York Times*, May 23, 1921; S. D. Stokes to James P. Woods, November 3, 1922, Stokes Papers, WVRHC. Not long after Stokes wrote to Woods asking that something be done about Smith, Smith lost his position as Stone Mountain superintendent. Assertion based on information from the 1922 *Annual Department of Mines*.

12　The two most prominent examples are Sid Hatfield and C. E. Lively. Even before his death, Sid Hatfield was viewed as either a folk hero or a bully "in a cheap snuff-colored suit." See: "Testimony of Mrs. Stella Scales," Lewis Collection, ERCA; and "Sid Hatfield Bows to Mingo Sheriff," *New York Times*, May 25, 1921; Contemporary newspaper descriptions of Lively depict him as a man of "gentle speech" and "suave manners": "Fed Miners to Get Truth About Battle," unknown newspaper, February 25, 1921, Matewan Omnibus Collection, ERCA. While Hatfield's stature as hero has grown, Lively has been vilified: Paul Lively interview. However, most of the project narrators had little difficulty acknowledging the duality of Hatfield's and other Massacre participants' natures. See the interviews of: Hawthorne Burgraff, Venchie Morrell (1990), and Bertha Damron; Ruby Aliff interview; Sean Wilentz, *Chants Democratic: New York City & The Rise of the American Working Class, 1788–1850* (New York: Oxford University Press, 1986 [1984]), 13.

13　In an interview with Michael Morrell, Herbert Gutman noted that David Montgomery's studies of American labor history have shown that certain eras in American history are noteworthy because they were hallmarks in the recurrent struggle over the meaning of America. The three cited—the 1830s, the 1880s, and the 1910s–1920s—were also pivotal decades in the history of the Tug Valley. "Herbert Gutman Interview," in Gutman, *Power and Culture: Essays on the American Working Class*, 329–356, 334; One scholar has noted that in "nine-tenths" of the counties across America, local political affairs have been controlled by "courthouse gangs." Lane W. Lancaster, *Government in Rural America* (New York: Van Nostrand, 1952), 57; For a social history of American industrialization, see Gutman, *Work, Culture and Society;* As historian Irving Bernstein has noted, violence is an "extraordinary American tradition of resolving disputes with guns instead of words," Bernstein, 43.

14   Venchie Morrell, 1989 and 1990 interviews; Sallie Dickens interview; Bertha Damron interview; Harold Dickens interview.

15   Manuel Barrios, interview with Rebecca Bailey, Summer 1989 Matewan Oral History Project.

16   Bertha Damron interview.

17   "Mingo County Has Highest Unemployment Rate in State," *Charleston Gazette*, August 24, 1999.

BIBLIOGRAPHY

ARCHIVAL COLLECTIONS
Catholic University, Washington, D.C.
John Mitchell Papers. Microfilm copy of this collection also available
in Microfilm Collection, Wise Library, West Virginia University.

Eastern Regional Coal Archives, Craft Memorial Library,
Bluefield, West Virginia
William Y. Cooper Collection
David E. Johnston Collection
H. C. Lewis Collection
Roland Luther Collection
Matewan Omnibus Collection
Pocahontas Land Corporation Collection
Ernie Reynolds Collection
Matewan Development Center, Matewan, West Virginia

Matewan Oral History Project
Dixie Accord, interview with John Hennen, Summer 1989.
"Smokey" Mose Adkins, interview with John Hennen, Summer
1989.
Ruby Aliff, interview with Rebecca J. Bailey, Summer 1989
Mattie Allara, interview with Rebecca J. Bailey, Summer 1989.
Jim Backus, interview with John Hennen, Summer 1989.
Manuel Barrios, interview with Rebecca Bailey, Summer 1989.
Harry Berman, interview with John Hennen, Summer 1989.
Vicie Blackburn, interview with Rebecca J. Bailey, Summer 1990.
Archie Bland, interview with Rebecca J. Bailey, Summer 1990.
Tom Blankenship, interview with C. Paul McAlister, Summer 1990.
Edith Boothe, interview with Rebecca J. Bailey, Summer 1989.

Hawthorne Burgraff, interview with John Hennen, Summer 1989.

Eva Cook, interview with Rebecca J. Bailey, Summer 1990.

James Curry, interview with Rebecca J. Bailey, Summer 1990.

Bertha Damron, interview with Rebecca J. Bailey, Summer 1989.

Harold Dickens, interview with Rebecca J. Bailey, Summer 1990.

Sallie Dickens, interview with Rebecca J. Bailey, Summer 1990.

Everett Faddis, interview with John Hennen, Summer 1989.

Johnny Fullen, interview with Rebecca J. Bailey, Summer 1990.

Virginia Grimmett, interview with Rebecca J. Bailey, Summer 1989.

Clarence "Dutch" Hatfield, interview with Rebecca J. Bailey, Summer 1989.

Ernest Hatfield, interview with John Hennen, Summer 1989.

Margaret Hatfield, interview with Rebecca J. Bailey, Summer 1990.

Josephine Hope, interview with Rebecca J. Bailey, Summer 1990.

Paul Lively, interview with Rebecca J. Bailey, Summer 1990.

John McCoy, interview with John Hennen, Summer 1989.

Venchie Morrell, interview with John Hennen, Summer 1989.

Venchie Morrell, interview with Rebecca J. Bailey, Summer 1990.

Daisy Nowlin, interview with Rebecca J. Bailey, Summer 1989.

Rose Nenni Ore, interview with John Hennen, Summer 1989.

Hiram Phillips, interview with John Hennen, Summer 1989.

Kenny Phillips, interview with John Hennen, Summer 1989.

Stella Pressley, interview with Rebecca J. Bailey, Summer 1989.

Harold (Howard) Radford, interview with John Hennen, Summer 1989.

Jeannette Simpkins, interview with Rebecca J. Bailey, Summer 1990.

Rufus Starr, interview with John Hennen, Summer 1989.

Bertha Staten, interview with Rebecca J. Bailey, Summer 1989.

Melvin Triolo, interview with John Hennen, Summer 1989.

Special Collections of the University of Virginia, Alderman Library, University of Virginia, Charlottesville, Virginia

E. L. Stone Papers

State Historical Society of Wisconsin, Madison, Wisconsin

The Papers of John L. Lewis, Microfilm Collection

West Virginia and Regional History Collection, West Virginia University, Morgantown, West Virginia

Manuscript Collections

Justus Collins Papers

John J. Cornwell Papers

Henry D. Hatfield Papers

Roy H. Keadle Papers

Matewan Massacre Papers

George C. McIntosh Papers

Ephraim F. Morgan Papers

Samuel Davis Stokes Papers

W. P. Tams, interview with Richard Hadsell. Typescript. *Although this is an oral history interview, it was transcribed and accessioned by the West Virginia and Regional History Collection into the Manuscript Collection.*

A. B. White Papers

Oral History Collection

William Carey, interview with Keith Dix, October 17, 1971, at Red Jacket, Mingo County, WV.

A. D. Lavinder, interview with Bill Taft and Lois McLean, June 22, 1973, at Matewan, Mingo County, WV.

Pamphlet Collection

Bias, B. Randolph "Condensed Facts: Relating to the Republican Organization in Mingo County, W.Va." n.p., n.d. Pamphlet 8662.

------. "'The Truth About West Virginia,' an address by Randolph Bias, an attorney of Williamson, W.Va., to the Kiwanis Club of Portland, Oregon, on September 12, 1922." Williamson, WV: n.p., 1923. Pamphlet 4344.

"'Confidential': Conservative Statement of the Situation of the Coal-Carrying Railroads, in their relation to the Coal and Coke Development of Kentucky, Virginia, West Virginia, and Tennessee (Compiled from an Official Report for 1903)." n.p., n.d. Pamphlet 7080.

"Governor Explains Why Federal Troops Brought Into West Virginia—Six Thousand Deluded Men In Insurrection, Reason He Offers." New York Commercial, 1921 [reprint]. Pamphlet 7640.

"Telegram from Harold W. Houston to the National Executive Committee, Socialist Party, read at the meeting of the committee, 11 May 1913," reported by Grace Silver in "National Committee Meeting," Socialist Party, in *International Socialist Review. Several sections of this issue, pertaining to the Paint Creek and Cabin Creek Strike, 1912–1913, were removed and accessioned into the WVRHC Pamphlet Collection as Pamphlet 191.*

West Virginia State Council of Defense. "Report of the Secretary on the Operation of the Compulsory Work Law, for the Year Ending June 19, 1918." Charleston, WV: Tribune Printing Company, 1918. Pamphlet 2902.

Young, Houston G., Secretary of State of West Virginia. "State of West Virginia: Official Returns of the General Election, held November 2, 1920." Charleston, WV: Tribune Printing Co, 1920. J. Hop Woods Bound Pamphlet Collection, 11476, volume 15.

Rare Book Collection

"A Virginia Boy," [pseudonym]. *Sodom and Gomorrah Today, or the History of Keystone, West Virginia,* n.p., 1912.

Sowers, A. D. *Some Facts about McDowell County, West Virginia.* Keystone, WV: A. D. Sowers, 1912.

UNPUBLISHED MATERIALS IN THE AUTHOR'S POSSESSION

Hatfield, Margaret. Undated correspondence with author, 1997–1998.

UNITED STATES DOCUMENTS

Bituminous Operators Special Committee. *United Mine Workers in West Virginia,* submitted to the United States Coal Commission, August 1923. n.p., 1923.

"Contested Election Case of *Rankin Wiley v. James A. Hughes* from the Fifth Congressional District of West Virginia." Sixty-Second

Congress, 2d session, Report No.1229 [Mr. Covington, from the Committee on Elections No.1 submitted the following report to accompany H.Res.703]. Washington: Government Printing Office, 1912.

"Extracts from Speech of Harold W. Houston at Holly Grove, on Paint Creek, August 14, 1912," in *Conditions in the Paint Creek District, West Virginia: Hearings Before a Subcommittee of the Committee on Education and Labor*, Part 3. Washington: GPO, 1913.

Houston, Harold W. *Brief on behalf of the United Mine Workers of America, before the committee on Education and Labor, United States Senate, in the matter of the investigation of violence in the coal fields of West Virginia and adjacent territory.* n.p., 1921.

Olmsted, Harry. *Statement of Harry Olmsted, Chairman of the Labor Committee of the Operators Association of the Williamson Field to the Senate Investigating Committee, July 14, 1921.* Washington: W.F. Roberts Co., 1921.

Operators' Association of the Williamson Field. *Statement Before the Subcommittee of the Committee on Education and Labor of the United States Senate.* n.p., 1921.

United States Bureau of the Census. *Twelfth Census: Population of the United States in 1900.* NAMP.

United States Bureau of the Census. *Thirteenth Census: Population of the United States in 1910.* NAMP.

United States Bureau of the Census. *Thirteenth Census: Abstract of the Census, with Supplement for West Virginia.* Washington, D.C.: GPO, 1913.

United States Bureau of the Census. *Fourteenth Census: Population of the United States in 1920.* NAMP.

United States Bureau of the Census. *Fourteenth Census: State Compendium: West Virginia.* Washington, D.C.: GPO, 1925.

U.S. Senate Committee on Education of Labor. *West Virginia Coal Fields: Hearings . . . to investigate the recent acts of violence in the coal fields of West Virginia and adjacent territory and the causes which led to the conditions which now exist in said territory.* Washington: GPO, 1921.

U.S. Senate Committee on Education and Welfare. *West Virginia Coal Fields: Personal Views of Senator Kenyon and views of Senators Sterling, Phipps, and Warren* . . . . Washington: GPO, 1922. (67th Congress. 2d Session Senate Report 457).

WEST VIRGINIA DOCUMENTS

West Virginia Department of Health. *Biennial Report of the West Virginia State Department of Health, 1917–1918.* Charleston: Tribune Printing Company, 1918.

West Virginia Department of Health. *Biennial Report of the West Virginia State Department of Health, 1919–1920.* Charleston: Tribune Printing Company, 1920.

West Virginia Department of Mines. *Annual Report of the West Virginia Department of Mines, 1891–1922.*

West Virginia Education Department. *West Virginia Education Directory, for the School Year 1917–1918,* edited by M. P. Shawkey. Charleston, WV: Tribune Printing Company, 1918.

West Virginia Legislature. *Acts of the West Virginia Legislature, for the year 1917.* Charleston, WV: Tribune Printing Company, 1917.

West Virginia Legislature. *Acts of the West Virginia Legislature.* 2nd Extraordinary Session. Charleston: Tribune Printing Company, 1917.

West Virginia Legislature. *Acts of the West Virginia Legislature, for the year 1921.* Charleston, WV: Tribune Printing Company, 1921.

West Virginia Legislature, John J. Cornwell and others. "West Virginia in The War," in *West Virginia Legislative Handbook and Manual and Official Register, for the year 1918,* compiled and edited by John T. Harris, clerk of the Senate, 775–948. Charleston, WV: The Tribune Printing Company, 1918.

West Virginia Legislature. *West Virginia Legislative Handbook and Manual and Official Register, for the year 1917,* compiled and edited by John T. Harris, clerk of the Senate. Charleston, WV: The Tribune Printing Company, 1917.

West Virginia Legislature. *West Virginia Legislative Handbook and Manual and Official Register, for the year 1920,* compiled and edited by John T. Harris, clerk of the Senate. Charleston, WV: The Tribune Printing Company, 1920.

West Virginia Legislature. "An Act establishing the County of Mingo," In *Acts of Legislature of West Virginia, Twenty-Second Regular Session.* Charleston: Moses W. Donnally, 1895.

West Virginia Legislature. "Concerning Deadly Weapons, Etc. (Red Man's Act)." In *Acts of the West Virginia Legislature, for the year 1882.* Wheeling, WV: W. J. Johnston, Public Printer, 1882.

West Virginia Legislature. "The Idleness and Vagrancy Act, Senate Bill #7." In *Acts of the West Virginia Legislature, Second Extraordinary Session*, May 14–26. Charleston, WV: Tribune Printing Company, 1917.

West Virginia Legislature. "Special Deputy Bill" House Bill #34." In *Acts of the West Virginia Legislature for the year 1917.* Charleston: Tribune Printing Company, 1917.

West Virginia Legislature. "Williamson Charter Bill, Senate Bill #199." In *Acts of the Legislature of the State of West Virginia*, 1915.

West Virginia Legislature. "Williamson Mining and Manufacturing Company. Reports of Incorporation." In *Acts of the West Virginia Legislature for 1891.* Charleston, WV: Moses W. Donnally, 1891.

Stewart, J. H., Commissioner of the State of West Virginia Department of Agriculture. *Third Biennial Report of the West Virginia Department of Agriculture, 1917–1918.* Charleston, WV: Tribune Printing Company, 1918.

West Virginia. "Ex Parte Lavinder, et al." In *Reports of the West Virginia State Supreme Court of Appeals*, 88 (February 22, 1921–June 24, 1921): 713–721.

West Virginia State Supreme Court of Appeals. "Ferrell v. Ferrell." In *Reports of the West Virginia Supreme Court of Appeals*, 53 (March 28, 1903–November 21, 1903): 515–524.

West Virginia State Supreme Court of Appeals. "Hatfield v. Allison." In *Reports of the West Virginia Supreme Court of Appeals*, 57 (January 24, 1905–April 25, 1905): 374–384.

West Virginia State Supreme Court of Appeals. "Thacker Coal & Coke Company v. Burke et al." In *Reports of the West Virginia*

Supreme Court of Appeals 59 (February 15, 1906 – April 24, 1906): 253–262.

West Virginia State Supreme Court of Appeals. "State ex rel. M. Z. White, Relator v. County Court of Mingo County." In *Reports of the West Virginia Supreme Court of Appeals* 86 (March 16– September 21 1920): 517–518.

West Virginia State Supreme Court of Appeals. "State of West Virginia v. Elias Hatfield." In *Reports of the West Virginia Supreme Court of Appeals* 48 (April 14, 1900–December 21, 1900): 561– 576.

West Virginia State Supreme Court of Appeals. "State of West Virginia v. J. S. McCoy." In *Reports of the West Virginia State Supreme Court of Appeals* 91 (April 25, 1922–October 17, 1922): 262–268.

West Virginia State Supreme Court of Appeals. "Williamson, et al v. County Court," "Hurst, et al v. Same," "Stafford, et al v. Same." In *Reports of the West Virginia Supreme Court of Appeals,* 56 (June 14, 1904): 38–43. *These cases were heard together.*

West Virginia State Supreme Court of Appeals. "Williamson v. Musick." In *Reports of the West Virginia Supreme Court of Appeals,* 60 (April 24, 1906–November 27, 1906): 58–75.

NEWSPAPERS

Bluefield Daily Telegraph, Special Industrial Edition 1896, 1908, 1920–1922

Bluefield Observer 1992

Charleston Advocate 1907

Charleston Evening Mail 1894

Charleston Gazette 1916, 1920, 1999

Huntington Advertiser 1891, 1902

Logan Banner 1921

Martinsburg Herald 1917

Mingo Republican 1911–1917, 1920

McDowell Recorder 1918, 1920

New York Times 1920–1921

West Virginia Federationist 1920–1921
Wheeling Intelligencer 1914
Williamson Daily News 1913–1920, 1982
Williamson Enterprise 1908

JOURNALS

Coal Age 1911, 1918, 1920
Coal Trade Journal 1897–1898, 1911, 1920–1922
National Coal Mining News 1921
United Mine Workers Journal 1901, 1920–1922

BOOKS, ARTICLES, AND THESES

Atkinson, George W. *Bench and Bar of West Virginia*. Charleston,
    WV: Virginia Law Book Company, 1919.

Atkinson, George W. and Alvaro F. Gibbens. *Prominent Men of West
    Virginia*. Wheeling, WV: W.L. Callin, 1890.

Ayers, Edward L. *The Promise of the New South: Life After
    Reconstruction*. New York: Oxford University Press, 1992.

Bagby, Wesley M. *The Road to Normalcy: the Presidential Campaign
    and Election of 1920*. Baltimore: Johns Hopkins Press, 1968
    [1962].

Bailey, Kenneth R. "A Judicious Mixture: Negroes and Immigrants
    in the West Virginia Mines, 1880–1917." *West Virginia History* 34
    (January 1973): 141–161.

-----. "A Temptation to Lawlessness: Peonage in West Virginia, 1903–
    1908." *West Virginia History* 50 (1991): 25–45.

Bailey, Rebecca J. "Matewan Before the Massacre: Politics, Coal, and
    the Roots of Conflict in Mingo County, 1793–1920." Ph.D. diss.,
    West Virginia University, 2001.

Barb, John M. "Strikes in the Southern West Virginia Coal Fields,
    1912–1922." Master's thesis, West Virginia University, 1949.

Barkey, Frederick A. "The Socialist Party in West Virginia from 1898
    to 1920: A Study in Working Class Radicalism." Ph.D. diss.,
    University of Pittsburgh, 1979.

Batteau, Allen W. *The Invention of Appalachia*. Tucson, AZ:

University of Arizona Press, 1990.

Beik, Mildred Allen. *The Miners of Windber: The Struggles of New Immigrants for Unionization, 1890s–1930*. University Park, PA: The Pennsylvania University Press, 1997 [1996].

Bernstein, Irving. *The Lean Years: A History of the American Worker, 1920–1933*. New York: Da Capo Press, 1983 [1960].

Berthoff, Rowland. *An Unsettled People: Social Order and Disorder in American History*. New York: Harper & Row, 1971.

Billings, Dwight B. and Kathleen M. Blee. *The Road to Poverty: The Making of Wealth and Hardship in Appalachia*. Cambridge: Cambridge University Press, 2000.

Blankenhorn, Heber. "Marching Through West Virginia." *Nation* 113 (September 14, 1921): 288.

Brown, Donna L. *Logan County Marriages, Book 1:1872–1892*. Bruno, WV: the author, n.d.

Burckel, Nicholas C. "Progressive Governors in the Border States: Reform Governors of Missouri, Kentucky, West Virginia, and Maryland, 1900–1918." Ph.D. diss., University of Wisconsin, 1971.

Burgett Richard. Interview. In *On Dark and Bloody Ground: An Oral History of the UMWA in Central Appalachia, 1920–1935*, edited by Anne Lawrence, 105–109. Charleston, WV: Miner's Voice, 1973.

Burkinshaw, Neil. "Labor's Valley Forge." *Nation* 111 (December 1920): 639.

Callahan, James M. *Semi-Centennial History of West Virginia*. Charleston, WV: Semi-Centennial Commission of West Virginia, 1913.

Campbell, John C. *Southern Highlander and His Homeland*. Lexington, KY: University Press of Kentucky, 1969 [reprint of 1921 edition published by the Russell Sage Foundation].

Carter, Charles Frederick. "Murder to Maintain Coal Monopoly." *Current History* 15 (1922): 597–603.

Chafin, Raymond. *Just Good Politics: The Life of Raymond Chafin, Appalachian Boss*. Pittsburgh: University of Pittsburgh Press, 1996.

Chapman, Mary Lucille. "The Influence of Coal in the Big Sandy
   Valley." Ph.D. diss., University of Kentucky, 1945.

Chernow, Ron. *The House of Morgan: An American Banking Dynasty
   and the Rise of Modern Finance.* New York: Simon & Schuster,
   1990.

*The Coal Catalog: Combined with a Coal Field Directory for the Year
   1920.* Pittsburgh, PA: Keystone Publishing Company, 1920.

Cole, Merle T. "Martial Law and Major Davis as 'Emperor of Tug
   River.'" *West Virginia History* 43 (Winter 1982): 118–144.

------. "The Department of Special Deputy Police, 1917–1919." *West
   Virginia History* 44 (Summer 1983): 321–333.

Coleman, McAlister. *Men and Coal.* New York: Farrar and Rinehart,
   1943.

Collier, James G. "Mingo County in World War II." Master's thesis,
   West Virginia University, 1956.

Conley, Phil M. *History of the West Virginia Coal Industry.* Charleston:
   Education Foundation Inc., 1960.

Conley, Phil M. "The Founder of the City of Williamson." *West
   Virginia Review* 2 (February 1925): 162.

Corbin, David A. "'Frank Keeney is Our Leader and We Shall Not
   Be Moved': Rank and File Leadership in the West Virginia Coal
   Fields." In *Essays in Southern Labor History: Selected Papers from
   the Southern Labor History Conference, 1976,* edited by Gary M.
   Fink and Merle E. Reed, *Contributions in Economics and Economic
   History,* no. 16, 144–156. Westport, CT: Greenwood Press, 1976.

------. *Life, Work, and Rebellion, The Southern West Virginia Miners,
   1880–1922.* Urbana: University of Illinois Press, 1981.

------. *The West Virginia Mine Wars: An Anthology.* Charleston, WV:
   Appalachian Editions, 1991.

Cornwell, John J. *A Mountain Trail: From the Farm to Schoolroom, to
   the Editor's Chair, the Lawyer's Office and the Governorship of West
   Virginia.* Philadelphia: Dorrance and Company Publishers, 1939.

Crowe-Carraco, Carol. *The Big Sandy.* Lexington, KY: University
   Press of Kentucky, 1979.

Cubby, Edwin A. "Railroad Building and the Rise of the Port of Huntington." *West Virginia History* 32 (October 1970): 234–247.

------. "Transformation of the Tug and Guyandot Valleys, Economic Development and Social Change in West Virginia, 1888–1921." Ph.D. diss., Syracuse University, 1962.

Dailey, Jane. "Deference and Violence in the Postbellum South: Manners and Massacres in Danville, Virginia." *Journal of Southern History* 63 (August 1997): 553–590.

Dillon, Lacy A. *They Died For King Coal.* Winona, MN: Apollo Books Inc., 1985.

Diner, Steven J. *A Very Different Age: Americans in the Progressive Era.* New York: Hill and Wang, 1998.

Dix, Keith, *What's a Miner to Do: The Mechanization of Coal Mining.* Pittsburgh: University of Pittsburgh Press, 1988.

------. *Work Relations in the Coal Industry: The Handloading Era, 1880–1930.* Morgantown, WV: Institute for Labor Studies, Division of Social and Economic Development, Center for Extension and Continuing Education, West Virginia University, 1989.

Doolittle, Edwin S. "On the Circuit in Southern West Virginia." *The Green Bag* 13 (1901): 284–286.

Dubofsky, Melvyn and Warren Van Tine. *John L. Lewis: A Biography.* New York: The New York Times Book Company, 1977.

Dunbar, John L. "Two Periods of Crisis in Labor-Management Relations in the West Virginia Coalfields, 1912–1913 and 1920–1922." Master's thesis, Columbia University, 1946.

Dunn, Durwood. *Cades Cove: The Life and Death of a Southern Appalachian Community, 1818–1937.* Knoxville: University of Tennessee Press, 1988.

Eller, Ronald D. *Miners, Millhands, and Mountaineers: Industrialization of the Appalachian South, 1880–1930.* Knoxville: University of Tennessee Press, 1982.

Ely, William. *The Big Sandy Valley: A History of the People and Country from the Earliest Settlement to the Present Time.*

Catlettsburg: Central Methodist Publishing, 1887.

Fagge, Roger. *Power, Culture and Conflict in the Coalfields: West Virginia and South Wales, 1900–1922*. New York: University of Manchester Press, 1996.

Fairchild, Herman L. *Memorial of Israel C. White*. n.p., 1928 [reprint from the *Bulletin of the Geological Society of America*, vol.39].

Fetherling, Dale. *Mother Jones, the Miners' Angel: A Portrait*. Carbondale, IL: Southern Illinois University Press, 1979.

Fisher, Lucy Lee, "John J. Cornwell, Governor of West Virginia, 1917–1921" *West Virginia History* 24 (April 1963/July 1963): 258–288, 370–389.

Foner, Philip S. *Women and the American Labor Movement: From Colonial Times to the Eve of World War I*. New York: The Free Press, 1980.

Franklin, John Hope and August Meier, eds. *Black Leaders of the Twentieth Century*. Urbana: University of Illinois Press, 1982.

Gaventa, John. *Power and Powerless: Quiescence and Rebellion in an Appalachian Valley*. Urbana: University of Illinois Press, 1980.

Glaab, Charles N. and A. Theodore Brown. *A History of Urban America*. New York: Macmillan, 1983 [1967].

Gleason, Arthur. *The Book of Arthur Gleason: My People*. New York: W. Morrow and Company, 1929.

------. "Company-Owned Americans," *Nation* 110 (12 June 1920): 794–795.

------. "Gunmen in West Virginia," *New Republic* 28 (21 September 1921): 90–92.

------. "Private Ownership of Public Officials," *Nation* 110 (May 1920): 724–725.

Gutman, Herbert. *Power and Culture: Essays on the American Working Class*, edited by Ira Berlin. New York: Pantheon Books, 1987.

------. *Work, Culture, and Society in Industrializing America: Essays in American Working Class and Social History*. New York: Vintage Books, 1976 [1966].

Hadsell, Richard M. and William E. Coffey. "From Law and Order to Class Warfare: Baldwin-Felts Detectives in the Southern West Virginia Coal Fields," *West Virginia History* 40 (Spring 1979): 268–286.

Hall, Jacquelyn Dowd, et al. *Like a Family: The Making of a Southern Cotton Mill World*. Chapel Hill and London: The University of North Carolina Press, 1987.

Harris, Evelyn L. K. and Frank J. Krebs, *From Humble Beginnings: West Virginia State Federation of Labor, 1903–1957*. Charleston, WV: Jones Printing Co., 1960.

Hartog, Hendrik. *Public Property and Private Power: The Corporation of the City of New York in American Law, 1730–1870*. Chapel Hill: University of North Carolina Press, 1983.

Hennen, John. *The Americanization of West Virginia: Creating a Modern Industrial State, 1916–1925*. Lexington, KY: University Press of Kentucky, 1996.

Hinrichs, A. F. *The United Mine Workers of America and the Non-Union Coalfields*. New York: n.p., 1923.

Hofstadter, Richard. *The Age of Reform: From Bryan to FDR*. New York: Alfred A. Knopf, 1956.

Howe, Barbara J. "West Virginia Women's Organizations, 1880s-1930 or 'Unsexed Termagants' . . . Help the World Along." *West Virginia History* 59 (1990): 81–102.

Hunt, Edward Eyre, F. G. Tryon, and Joseph H. Willits., eds. *What The Coal Commission Found: An Authoritative Summary By The Staff*. Baltimore, MD: The Williams & Wilkins Company, 1925.

Hurst, Mary. "A Social History of Logan County, West Virginia, 1765–1928." Master's thesis, Columbia University, 1928.

Ireland, Robert M. *Little Kingdoms: The Counties of Kentucky, 1850–1891*. Lexington: University Press of Kentucky, 1977.

Jeffreys-Jones, Rhodri. *Violence and Reform in American History*. New York: Franklin Watts, 1978.

Johnson, James P. *The Politics of Soft Coal: The Bituminous Industry from World War I through the New Deal*. Urbana: University of Illinois Press, 1979.

Jones, Virgil Carrington. *The Hatfields and the McCoys.* New York: Ballantine Books, 1948.

Jordan, Daniel P. "The Mingo War: Labor Violence in the Southern West Virginia Coal Fields, 1919–1922." In *Essays in Southern Labor History; Selected Papers, Southern Labor History Conference, 1976,* edited by Gary M. Fink and Merl E. Reed, *Contributions in Economics and Economic History,* no. 16, 102–143. Westport, CT: Greenwood Press, 1977.

Kane, Anne. "Reconstructing Culture in Historical Explanation: Narratives as Cultural Structure and Practice." *History and Theory* 39 (October 2000): 311–330.

Karr, Carolyn. "A Political Biography of Henry Hatfield." *West Virginia History* 28 (October 1966/January 1967): 35–63, 137–140.

Kirk, John W., compiler. *Progressive West Virginians.* Wheeling: Wheeling Intelligencer, 1923.

Klingaman, William K. *1919: The Year Our World Began.* New York: Harper & Row, 1989 [1987].

Kornweibel, Theodore, Jr., ed. "Reports by Informant C-61 to A. E. Hayes, for the Southern District of West Virginia, 9 October–30 October 1920." In *Federal Surveillance of Afro-Americans (1917–1925): The First World War, the Red Scare, and the Garvey Movement* (microfilm project of University Publications of America).

Lambie, Joseph T. *From Mine to Market: The History of Coal Transportation on the Norfolk and Western Railway.* New York: New York University Press, 1954.

Lancaster, Lane W. *Government in Rural America.* New York: Van Nostrand, 1952.

Lane, Winthrop D. *Civil War in West Virginia: A Story of the Industrial Conflict in the Coal Mines.* New York: B.W. Huebsch, 1921 [reprint 1994].

Laslett, John H. M. *Colliers Across the Sea: A Comparative Study of Class Formation in Scotland and the American Midwest, 1830–1924.* Urbana: University of Illinois Press, 2000.

Laurie, Clayton. "The United States Army, The Return to Normalcy in Labor Dispute Interventions: The Case of the West Virginia Coal Mine Wars, 1920–1921." *West Virginia History* 50 (1991): 1–24.

Lawson, Sidney B. *Fifty Years a Country Doctor: Autobiography and Reminiscences of Sidney B. Lawson, M.D.* Logan, WV: n.p., 1941.

Lee, Howard B. *Bloodletting in Appalachia: The Story of West Virginia's Four Major Mine Wars and Other Thrilling Incidents of Its Coal Fields.* Morgantown, WV: West Virginia University Library, 1969.

Lewis, Ronald L. *Black Coal Miners in America: Race, Class, and Community Conflict, 1780–1980.* Lexington, KY: University of Kentucky Press, 1987.

------. *Transforming the Appalachian Countryside: Railroads, Deforestation, and Social Change in West Virginia, 1880–1920.* Chapel Hill: University of North Carolina Press, 1998.

Lunt, Richard D. *Law and Order vs. The Miners: West Virginia, 1906–1933.* Charleston, WV: Appalachian Editions, 1992.

McDowell County DAR. *McDowell County.* Fort Worth, TX: University Supply and Equipment Company, 1959.

McKinney, Gordon B. "Industrialization and Violence in Appalachia in the 1890s," *An Appalachian Symposium: Essays Written in Honor of Cratis Williams*, edited by J. W. Williamson, 131–144. Boone, NC: Appalachian State University Press, 1977.

McWhorter, J. C. "Abolish the Jury." *West Virginia Law Quarterly* 29 (January 1923): 97–108.

Montell, William Lynwood. *Killings: Folk Justice in the Upper South.* Lexington: University Press of Kentucky, 1986.

Mooney, Fred. *Struggle in the Coal Fields: The Autobiography of Fred Mooney.* Morgantown, WV: West Virginia University Library, 1967.

Munn, Robert F. "The Development of Model Company Towns in the Bituminous Coal Fields." *West Virginia History* 40 (Spring 1979): 243–253.

Murray, Robert K. *Red Scare: A Study of National Hysteria, 1919–1920.* New York: McGraw Hill, 1964.

Murphy, Paul L. *World War I and the Origins of Civil Liberties in the United States.* New York: Norton, 1979.

Murphy, Robert E., compiler. *Progressive West Virginians: Some of the Men Who Have Built Up and Developed the State of West Virginia.* Wheeling, WV: The Wheeling News, 1905.

Nash, Michael. *Conflict and Accommodation: Coal Miners, Steel Workers, and Socialism, 1890–1920.* Westport, CT: Greenwood Press, 1982.

Nisbett, Richard E. and Dov Cohen. *Culture of Honor: The Psychology of Violence in the South.* Boulder, CO and Oxford: Westview Press, 1996.

Palladino, Grace. *Another Civil War: Labor, Capital, and the State in the Anthracite Regions of Pennsylvania, 1840–1868.* Urbana and Chicago: University of Illinois Press, 1990.

Payne, Henry M. "The Future of Williamson and the Tug River Coal Field," *The Illustrated Monthly West Virginian* 7 (August 1908): 45–49.

Penn, Neil Shaw. "Henry D. Hatfield and Reform Politics: A Study of West Virginia Politics from 1908 to 1917." Ph.D. diss., Emory University, 1977.

Perry, Huey. *They'll Cut Off Your Project: A Mingo County Chronicle.* New York: Praeger Press, 1972.

Phillips, Cabell. "The West Virginia Mine War." *American Heritage* (August 1974): 58–61, 90–96.

Posey, Thomas E. "Some Significant Aspects of the West Virginia Labor Movement." *West Virginia Academy of Science Proceedings* 22, West Virginia University *Bulletin,* ser. 51, no.12-3 (June 1951): 120–127.

Rakes, Paul H. "Technology in Transition: The Dilemmas of Early Twentieth Century Coal Mining." *Journal of Appalachian Studies* 5 (Spring 1999): 27–60.

Rice, Bradley Robert. *Progressive Cities: The Commission Government Movement in America, 1901–1920.* Austin, TX: University of Texas, 1977.

Rice, Otis K. *The Hatfields and The McCoys.* Lexington, KY: University of Kentucky Press, 1982.

Rice, Otis K. and Stephen W. Brown. *West Virginia: A History.* 2d ed. Lexington, KY: University Press of Kentucky, 1993.

Ross, Malcolm. *Death of a Yale Man.* New York: Farrar and Rinehart, 1939.

------. *Machine Age in the Hills.* New York: The MacMillan Co., 1983.

Roy, Andrew. *A History of the Coal Miners of the United States.* Columbus, OH: Press of J. L. Trauger Printing Co, 1902.

------. "The Thacker Coal Field of West Virginia." *Mines and Minerals* 19 (May 1899): 472.

Said, Edward W. *Orientalism.* New York: Pantheon Books, 1978.

Salstrom, Paul. *Appalachia's Path to Dependency: Rethinking a Region's Economic History, 1730–1940.* Lexington, KY: University Press of Kentucky, 1994.

------. "Newer Appalachia as One of America's Last Frontiers." In *Appalachia in the Making: The Mountain South in the 19th Century,* edited by Dwight Billings, Mary Beth Pudup, and Altina Waller, 76–102. Chapel Hill: University of North Carolina Press, 1995.

Seltzer, Curtis. *Fire in the Hole: Miners and Managers in the American Coal Industry.* Lexington, KY: University Press of Kentucky, 1985.

Shapiro, Henry D. *Appalachia On Our Mind: The Southern Mountains and Mountaineer in the American Consciousness, 1870–1920.* Chapel Hill: University of North Carolina Press, 1978.

Shelton, Floyd Bunyon. "An Investigation of the Social Life of a West Virginia Coal Field." Bachelor's thesis, Emory University, 1920.

Shifflett, Crandall. *Coal Towns: Life, Work and Culture in Company Towns in Southern Appalachia, 1880–1960.* Knoxville: University of Tennessee Press, 1994.

Shinedling, Abraham J. *West Virginia Jewry: Origins and History*, 3 vols. Philadelphia: Maurice Jacobs, Inc, 1963.

Skeen, David O. "Industrial Democracy, Social Equality, and Violence: The West Virginia Mine Wars: 1912–1921." Masters' thesis, California State University, 1996.

Smith, Nancy Sue. *An Early History of Mingo County, West Virginia*. Williamson, WV: Williamson Printing Co., 1960.

Soule, George H. and Vincent P. Carosso. *American Economic History*. New York: Dryden Press, 1957.

Spence, Robert Y. *Land of the Guyandot: A History of Logan County*. Detroit: Harlo Press, 1976.

Spivak, John L. *A Man in His Time*. New York: Horizon Press, 1967.

Stock, Catherine McNicol. *Rural Radicals: Righteous Rage in the American Grain*. Ithaca, NY: Cornell University Press, 1996.

Striplin, E. F. *The Norfolk and Western: A History*. Roanoke, VA: Norfolk and Western Railway Company, 1981.

Sullivan, Kenneth. "Coal Men of the Smokeless Coalfields." *West Virginia History* 41 (Winter 1980): 142–165.

Thelen, David. *Paths of Resistance: Tradition and Dignity in Industrializing Missouri*. Columbia: University of Missouri Press, 1991 [reprint New York: Oxford University Press, 1986].

Thomas, Jerry Bruce. "Coal Country: The Rise of the Southern Smokeless Coal Industry and Its Effect on Area Development, 1872–1910." Ph.D. diss., University of North Carolina, 1971.

Thompson, E. P. *Customs in Common*. New York: New Press: Distributed by W.W. Norton, 1991.

------.*The Making of the English Working Class*. New York: Vintage Books, 1966 [1963].

Thompson, Paul. *Voice of the Past: Oral History*, third edition. New York: Oxford University Press, 2000.

Thurmond, Walter R. *The Logan Coal Field of West Virginia: A Brief History*. Morgantown, WV: West Virginia University Library, 1964.

Titler, George J. *Hell in Harlan.* Beckley, WV: BJW Printers, n.d.

Trail, William R. "The History of the United Mine Workers in West Virginia, 1920–1945." Master's thesis, New York University, 1950.

Trotter, Joe William. "Black Miners in West Virginia: Class and Community Responses to Workplace Discrimination, 1920–1930." In *The United Mine Workers of America: A Model of Industrial Solidarity?* edited by John H.M. Laslett, 269–296. University Park, PA: The Pennsylvania State University Press, 1996.

------. *Coal, Class, and Color: Blacks in Southern West Virginia, 1915–1932.* Urbana: University of Illinois Press, 1990.

Tucker, Gary J. "William E. Glasscock and the Election of 1910," *West Virginia History* 40 (Spring 1979): 254–267.

Tudiver, Sara Lubitsch. "Political Economy and Culture in Central Appalachia, 1790–1977." Ph.D. diss., University of Michigan, 1984.

Turner, William P. "From Bourbon to Liberal: The Life and Times of John T. McGraw, 1856–1920." Ph.D. diss., West Virginia University, 1960.

Velke, John A. *Baldwin-Felts Detectives, Inc.* Richmond, VA: s.n., 1997.

Vinson, Z. T. (Zachary Taylor). "Advocating Co-operation and Organization of West Virginia's Coal Operation," An address before the West Virginia Mining Institute 10 December 1914, in Huntington WV. n.p., n.d.

------. "Railway Corporations and the Juries." *Minutes of the West Virginia Bar Association, 17th Annual Meeting.* Clarksburg, February 12–13, 1902: 42–51.

Wallace, George S. *Cabell County Annals and Families.* Richmond: Garrett & Massie, 1935.

Waller, Altina L. *Feud: Hatfields, McCoys and Social Change in Appalachia, 1860–1900.* Chapel Hill: University of North Carolina Press, 1988.

------. "Feuding in Appalachia: The Evolution of a Cultural Stereotype." In *Appalachia in the Making: The Mountain South in the 19th Century*, edited by Dwight Billings, Mary Beth Pudup, and Altina Waller, 347–376. Chapel Hill: University of North Carolina Press, 1995.

Walls, Emick R. "West Virginia's Greatest Manhunt." *West Virginia Illustrated* 2 (May–June 1971): 29–33.

Warner, Arthur. "West Virginia: Industrialism Gone Mad." *Nation* 113 (October 1921): 372–373.

-----. "Fighting Unionism with Martial Law." *Nation* 113 (October 1921): 395–396.

Washington, James. Interview. In *On Dark and Bloody Ground: An Oral History of the UMWA in Central Appalachia, 1920–1935*, edited by Anne Lawrence, 107–109. Charleston, WV: Miner's Voice, 1973.

West Virginia Heritage Encyclopedia. Edited by Jim F. Comstock. 25 vols. Richwood, WV: Jim Comstock, 1976.

West Virginia State Gazetteer and Business Directory, 1891–1892, vol 4. Detroit: R. L.Polk, 1892.

West Virginia State Gazetteer and Business Directory, 1895–1896, vol. 5. Detroit: R. L.Polk, 1896.

West Virginia State Gazetteer and Business Directory, 1902–1903, vol. 8. Detroit: R. L.Polk, 1903.

Whisnant, David E. *Modernizing the Mountaineer: People, Power and Planning in Appalachia*. Knoxville: University of Tennessee Press, 1994.

Wiebe, Robert H. *The Search for Order: 1877–1920*. New York: Hill and Wang, 1967.

Wilentz, Sean. *Chants Democratic: New York City & The Rise of the American Working Class, 1788–1850*. New York: Oxford University Press, 1986 [1984].

Williams, John A. *West Virginia and the Captains of Industry*. Morgantown: West Virginia University Libraries, 1976.

-----. "The New Dominion and the Old: Antebellum and Statehood

Politics as the Background of West Virginia's 'Bourbon Democracy.'" *West Virginia History* 33 (July 1972): 317–407.

Wilson, Edmund. "Frank Keeney's Coal Diggers" *New Republic* 67 (8 and 15 July 1921): 195–199, 229–231.

Women's Christian Temperance Union of West Virginia. *Report of the Twenty-Sixth Annual Meeting of the Women's Christian Temperance Union of West Virginia, held at Huntington, WV, October 2–6, 1908*, edited by Mrs. K. M. Murill. Charleston, WV: Tribune Printing Company, 1909.

------. *West Virginia Woman's Christian Temperance Union, Twenty-Eighth Year, Charleston, WV, October 5, 6, 7, 1910*. Fairmont, WV: Index Print, n.d.

INDEX